# CHICAGO RIVER BRIDGES

CHICAGO RIV

# ER BRIDGES

Patrick T. McBriarty

University of Illinois Press ■ Urbana, Chicago, and Springfield

Photography by Laura Banick, Kevin Keeley,
Patrick McBriarty, and David M. Solzman.

Library of Congress Cataloging-in-Publication Data
McBriarty, Patrick T.
Chicago River bridges / Patrick T. McBriarty.
pages  cm
Includes bibliographical references and index.
ISBN 978-0-252-03786-3 (hardcover : alk. paper)
1. Movable bridges—Illinois—Chicago.
2. Movable bridges—Illinois—Chicago River.
I. Title.
TG420.M28     2013
388.4'11—dc23     2013003508

Dedicated to Barb Fox

# CONTENTS

# PREFACE

The genesis of this book is my appreciation for manufacturing, engineering, and industrial landscapes. I have always had a curiosity about how things work. This, combined with an affinity for the power, wonder, and gritty aesthetic inherent in heavy industry, aroused my interest in industrial structures. It was clear to me upon discovering Chicago's drawbridges that they held an inherent beauty that should be shared with a wider audience.

The seed was planted twelve years ago, one Sunday afternoon, when I wandered down to Kinzie Street with a camera and photographed the bridges. However, everyday life soon buried the idea. Six years later, while taking a year off from work, the seed germinated. Those twelve months soon became thirty-six, as researching the "bridge book" grew from an idea into a compulsion and productive obsession.

My early research uncovered a vast untapped history scattered across a variety of sources that was begging to be organized. I also quickly discovered that, amazingly, a complete book on Chicago's bridges did not exist. Both excited and scared about what I was getting into, the magnitude conveyed by these magnificent, silent steel structures drew me deeper into the topic. More than just a means to cross the river, the city's drawbridges offered a connection to Chicago's past amid the constantly changing urban landscape. The topic was irresistible, representing pragmatism, ingenuity, and Chicago's unique contribution to the evolution of moveable-bridge technology.

Originally interested in putting together a complete Chicago-bridge bible, it soon became clear that capturing the full history of all the bridges over Chicago's waterways would be too broad. With the Calumet and Chicago rivers, not to mention the canals, the research alone would add years of work. There was enough material, as good friend Phil Keeley pointed out, for two or three books. So I begrudgingly decided to go after the core of the history and focus on the Chicago River.

I discovered that even limiting myself to the Chicago River presented too broad a scope. The alterations to canals off the river included the Illinois & Michigan Canal, Sanitary & Ship Canal, and North Shore Canal just within the city limits. So I omitted the canals and nonnavigable portions of the North and South branches, which primarily contained fixed bridges and seemed tangential to the emerging moveable-bridge theme. I also elected to omit coverage of the railroad bridges, which entailed an entirely new vein of research and seemed best presented in a separate collection. I hope these omitted and significant chunks of Chicago's bridge history will be captured in some future work. Ultimately, this book was limited to the navigable portions of the

Chicago River; centered on Chicago's downtown, it presents the bridges of the Main Channel, the South Branch down to South Ashland Avenue, and the North Branch up to Belmont Avenue. The fifty-five bridge chapters contained herein detail the history of three tunnels and 173 Chicago River bridges, past and present.

It should be noted that throughout this book, I have chosen to use the less common spelling of *moveable* instead of *movable* to emphasize the element of movement inherent to the drawbridges. Similarly, the word *city* is used to refer to Chicago in its most general sense, while the capitalized word *City* refers to the municipal authority or more explicit "City of Chicago."

The book is organized to first introduce Chicago's general bridge history, moving to an explanation of the development of bridge designs and then, finally, to the core of the book containing the individual bridge chapters. The chapters are presented geographically in three parts: the Main Channel, South Branch, and North Branch of the Chicago River. Each bridge chapter starts with the current bridge (if one exists), followed by a chronology from the first bridge to the last bridge removed from that location.

Seeking to understand these structures and give the bridges their due over the past seven years has been a satisfying, frustrating, and consuming effort. I hope that this work will provide an appropriate guide and tribute to the magnificent Chicago River bridges and the individuals who designed, built, and worked on them.

# ACKNOWLEDGMENTS

t is hard to express all my thanks to the many people who have provided assistance, advice, editing comments, and support throughout the process of writing and completing this book. The repeated encouragement of my friends and queries of "How's the book coming?" have been greatly appreciated, despite my initial anguish at still being in the throes of completing this work.

For believing in the project and bringing it to fruition, I must thank the folks at the University of Illinois Press and in particular the chief acquisitions editor, Laurie Matheson. Her excitement and belief in the project have been invaluable in making this book a reality. Additionally, three peer-review readers provided thoughtful and critical suggestions that improved the manuscript; their input is appreciated.

Thanks go to Phil Keeley, who provided great understanding, editorial help, advice, and encouragement throughout the project. Similarly, my parents, Charlie and Eileen McBriarty, have been most supportive and helped edit the manuscript at key stages. Similar thanks go to James Scanlon and Paul Mendelson for their line editing.

For marine support, I must first thank Tom Neill for the use of his powerboat for the first photographic excursion on the river. I wish he were still here to see the results. Mary and Tim Corkell deserve a big thank-you for letting me keep their dinghy on the Chicago River for photographic trips between February and May 2007. Sam Sproviero allowed me to keep the dinghy and granted me access to the dock behind Green Dolphin Street. I want to thank my brother, Matt McBriarty, for use of his garage the past few winters to store the powerboat purchased for the project and also for his and Dave Ritchie's help on the numerous projects to fix and maintain this "pig of a boat." Thanks, guys, for the lipstick! Rick Hayslip, the manager at the Goose Island Boatyard, runs a solid operation. Doug MacFarlane repeatedly lent his truck to move the boat, spring and fall.

David Solzman, author of *The Chicago River* and professor of geography, deserves significant credit for cheerleading, guidance, and support at key points in the development of this book. I am also very thankful to him for allowing me to use the *List of City Bridges, 1914,* a bridge inventory he received from the estate and private collection of Richard Sutphin. Similarly, Paul Schellinger, the editorial director of reference at the University of Chicago Press, provided significant advice regarding the audience and voice for this book. The help of architects Jay Muller and Terry Sullivan of Muller+Muller, Ltd., and Brian Steele of the Chicago Department of Transportation has been greatly appreciated. Brian was instrumental in getting behind the scenes for several bridge lifts and putting me in contact with CDOT supervisor Daryl Rouse. Mr. Rouse has provided great insight and a valuable resource on Chicago bridge-lift operations.

Ulrich Danckers and Jane Meredith, authors of *Early Chicago,* have my thanks for meeting with me and reading and providing comments on an early manuscript. I cannot say enough about the value of their book on Chicago's beginnings and early history, peoples, and events.

Many others helped in a variety of ways: Dick Simpson, professor and former alderman, on the changing forms of government financing during Chicago's history; Robert Graham, archivist at the Historical Collections of the Great Lakes at Bowling Green State University, with several historical images; Richard Kahan, John Fincher, and Jack Dudley, for valuable insights into book publishing, sales, and distribution; Tom Lenard, for valuable background and history on Grebe's Boatyard; and Dave Cherry, regarding the history and demise of Chicago's traction system. Thanks also to the Bridge Department within the Chicago Department of Transportation for access to its archives and specifically to engineer Jay Orlando with Lake Shore Drive's Ogden Slip single-leaf bascule.

I am indebted to Grant Crowley, owner of Crowley's Yacht Yard, for his repeated help, advice, knowledge, and referrals to additional experts and for facilitating access to the Chicago Mar-

itime Museum archives. Additional thanks go to Dean Tank and, particularly, Don Glasell for their help at the Chicago Maritime Museum. Rick Strilky has been very helpful with referrals, support, and suggestions. Adrienne "Nikki" Guyer deserves thanks for her encouraging comments on an early draft of the book. Promotional ideas for the book from Harvey Moshman, a producer at Channel 2 News in Chicago, are also greatly appreciated.

Not least, photographers Kevin Keeley and Laura Banick deserve great thanks for their hard work, time, and excitement for this project. I am greatly indebted to Kevin in particular for his efforts and time. Kevin, I owe you many drinks, on which I am sure you will have many opportunities to collect. Thanks!

This book would not have been possible without the libraries and archives that hold, preserve, and maintain so many wonderful books, periodicals, images, and manuscripts for public use. I am immensely grateful to these institutions and their librarians, archivists, and administrators. Specific thanks go to the people at the Chicago Public Library, particularly the Government Collections' reference librarians and archivists at the Special Collections Archive. Likewise, thanks are due to the Chicago History Museum's Research Center librarians and archivists, especially Debbie Vaughan, who provided cheerful assistance so many times throughout my many visits there. Dan Wendt,

Peggy Bradley, and Gerald Austiff at the Metropolitan Water Reclamation District of Greater Chicago helped me with the historical images collection and gave me access to the library. Thanks to the librarians and archivists at the Newberry Library who were so helpful and professional and to the student aides and librarians at the Illinois Regional Archives Collection at the Northeastern University Library. I should also like to thank Kay Geary, the public services librarian at the Transportation Library of Northwestern University, for her help and excitement about the project. Thanks, everyone.

# CHICAGO RIVER BRIDGES

# INTRODUCTION

There is something magical about drawbridges. The simple idea of a moving bridge sparks the imagination. The paradox of a firm roadway pulling apart and opening conjures both danger and excitement. Using simple mechanics, moveable bridges *draw* our attention and incite a childlike fascination. In the long history of bridges, drawbridges are a relatively new and intriguing human invention.

Natural bridges were formed and destroyed long before the existence of humankind. Shifting continents, earthquakes, and volcanic eruptions reshaped the earth; wind and water carved and eroded the landscape. As life on earth flourished, a fallen tree across a stream allowed the deft of claw or foot to move from one side to the other. Bridges are a natural, ancient, and elemental concept.

Early peoples found great power in bridges, not just for transportation, but as a concept. A bridge as a metaphor is prevalent across many languages and cultures, for the notion of a bridge is uplifting and positive. It takes us places that were once forbidden, impossible, or inconceivable. A simple concept, bridges are priceless, connecting people, places, and things, literally and figuratively.

On a planet that is two-thirds water, the landmasses are separated by oceans and subdivided by rivers, lakes, and streams; thus, bridges become essential. For early humans, crossing these obstacles was dangerous or nearly impossible. Getting swept away, bashed among the rocks, and drowned or simply getting wet and cold while crossing a stream could result in sickness or even death. One alternative was traveling upstream in search of a natural ford to wade across, but natural bridges or fords were rarely where people wanted or needed them.

The development of tools allowed humans to evolve. Archaeological studies date the use of simple tools for crushing or killing food as early as 3.4 million years ago. However, recorded history reaches a mere 5,000 years into the past. We can only guess when the first bridge was built by humans. A simple bridge requires only a few basic tools and the idea to set the form in place. Ages ago, across isolated cultures and great distances, bridge building became universal.

With the advance of human society, bridges became political, strategic, and often military structures that allowed travel by soldiers and civilians alike. Before the invention of the wheel—and then because of it—bridges expanded to better spread civilization, conquest, and culture. These great utilitarian structures are among the earliest public works. Roman roads and bridges, for example, many still in existence today, were key in expanding the empire.

Like paving stones on a road, bridges are elemental building blocks of human advancement. Very few concepts are more basic to transportation than bridges. They have evolved, variations on a theme, as bigger, better, and more expansive structures conquering ever greater challenges. Yet even the simplest forms remain in common use today.

Bridges exist and have flourished because of the tremendous benefit and use they have provided throughout the ages. In most cases, investment in a bridge yields a value many times greater than its cost of construction. Bridges provide an easy path, save time and energy, and increase human productivity. They encourage commerce and offer immeasurable benefits on personal, economic, and cultural levels. The immediate value of a bridge compounds over time and frequently supports new or more intensive uses than the original builders could have imagined.

All natural bridges and most constructed bridges are fixed. A fixed bridge is the oldest and most common type of bridge. It is permanent and connects point A to point B across some obstacle. The evolution of fixed bridges has stretched humankind's ingenuity, technology, and engineering savvy. The biggest, highest, and longest bridges in the world are all fixed bridges. Such bridges require minimal maintenance, fit a variety of applications, are well suited for spanning great distances, last a long time, and are generally safe.

The moveable bridge is the other basic type. The term sometimes refers to a portable bridge used, often by the military, to quickly cross an obstacle by deploying, using, dismantling, moving, and reusing the bridge. For this book, moveable bridges are those that are fixed in place, moving only to open or close. The term *moveable bridge* is then generally interchangeable with what most people call a drawbridge.

The drawbridge is a uniquely human invention combining the concepts of connection and movement at a fixed location. Through man's ingenuity, the moveable bridge has evolved to offer a great variety of designs and uses to solve two basic applications for defense or the crossing of navigable waterways.

The earliest recorded evidence of drawbridges indicates their use for defensive purposes. Around 2000 BC, an Egyptian fortress at Buhen had two drawbridges to protect each entrance. These bridges could be drawn back on rollers to defend the fort from attack. During the Middle Ages, drawbridges were commonly used to protect the front gates of the growing number of keeps and castles. As larger and more elaborate defenses were built, the entrances also grew in size. Soon, drawbridges light enough to be moved by just a few men were inadequate.

Counterweights were employed to balance against the heavy wood doors, making bigger and heavier drawbridges practical. This ancient design became known as a bascule bridge and was the key to developing today's modern drawbridges.

*Bascule* is a French term that refers to a seesaw-like mechanical device. It is derived from *baculer,* which means to strike on the buttocks and likely originally described landing on one's buttocks (*bas* is the French equivalent of *down; -culer* is a verbal derivative of *rump* or *buttocks*). As shown by the diagrams, if the counterweights are the equivalent of the buttocks, opening such a bridge would cause it to bascule.

Castle bridges and most modern bascule bridges operate in the same manner, like a huge seesaw. The pivot point or point of rotation is typically located off center to create a short and long arm for the bridge. The long arm extends over the obstacle, usually a waterway; the short arm holds the counterweight to balance against the heavier long arm. The counterweight moves the bridge's center of gravity to or very near the bridge's point of rotation. The counterweight provides the mechanical advantage that allows heavy drawbridges to be operated with very little effort. The bascule bridge was used extensively for defense from the ninth into the fourteenth centuries.

Moveable bridges are not just for defense, though; they also solved the problem of crossing navigable waterways. Most cities develop around natural break points in transportation, such as a harbor, river mouth, or the junction of two rivers. As communities developed, conflict arose between land and water travel. The growing population centers needed reliable means for crossing the waterways, but fixed bridges potentially blocked the very shipping routes that created the cities. A high-clearance fixed bridge was often impossible or impractical; a moveable bridge

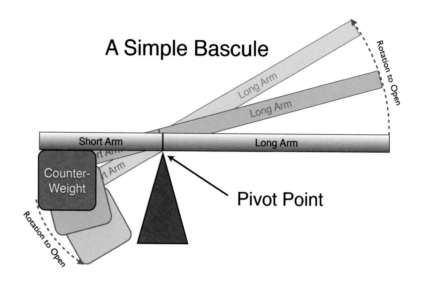

A Simple Bascule

The simplified bascule principle used in the design of most Chicago drawbridges.
© Patrick McBriarty

was the only reasonable solution that allowed both land and water travel. As early as 1275 BC, during the reign of Ramses II of Egypt, pontoon bridges crossed the Nile and were drawn open for the passage of ships.

Further growth and development of cities increased demand for bridges and new bridge forms like the arch, suspension, and bascule bridges evolved. The Industrial Revolution stimulated great innovations across areas of everyday life and brought material science and formal engineering to the vanguard of bridge building. Meanwhile, major infrastructural projects such as canals and railroads spurred further bridge design and innovation.

For many cities in an era of sailing ships and then steamers where water was the primary means of transportation, the development of fixed and moveable bridges over waterways was critical to continued growth. Nowhere was the development of the moveable bridge more important than in Chicago. Though not much more than thirty miles in length, the heavily trafficked Chicago River has had more drawbridges than any other river. The relatively young City of Chicago incorporated in 1837 and quickly transformed from a swampy outpost into a world-renowned city. Experiencing some of the most rapid population growth in history, Chicago became the busiest port in the world during the 1880s. The waterway cutting through the heart of downtown Chicago was soon stitched together almost exclusively by drawbridges. The drawbridges were a key element in fostering growth and creating a Chicago that would be unimaginable without its bridges.

In Chicago's first hundred years, the demands for free navigation of ships and the connection of street traffic across the river made the bridges a focal point for tension and conflict. They became the scenes of collision, drowning, flood, fire, and negligence that caused the destruction of bridges, damage to ships, and loss of life. Construction or removal of a bridge directly influenced the commerce of the city, generating controversy and strife. Local political battles were common, and eventually the Chicago River was placed under federal oversight. These forces forged many important bridging solutions. Though specific to Chicago, such solutions were frequently applicable to the bridging needs of the world.

Chicago's hundreds of bridges have supported the traffic of the city while also yielding right-of-way to ships bringing trade and commerce to the region. Today's steel and concrete structures spanning the waterways, some built at the beginning of the twentieth century, have opened and closed hundreds of thousands of times. Thousands of tons of roadway are lifted and then fitted back together gently and precisely. Operating to within fractions of an inch, these great machines have repeated this dance capably and effectively for decades and, in some cases, for a century or more. Chicago has long hosted the world's greatest collection of drawbridges.

How do these huge, aging machines work, and how did Chicago come to build the greatest network of moveable bridges of any municipality? The answer to this question is the fascinating story of Chicago's river bridges.

## Why Drawbridges Made Sense for Chicago

The development of Chicago's bridges parallels the development of the city itself. From the beginning, the Chicago River was an important waterway for Native Americans, then explorers, traders, and settlers, and connected them to western lands. In 1673 Jacques Marquette and Louis Jolliet were the first recorded Europeans to portage into the Chicago River and view the swampy land that today bears a metropolis. Starting at St. Ignace (now part of the state of Michigan), they navigated by canoe down Lake Michigan into Green Bay, up the Upper Fox River to Lake Winnebago, and down the Lower Fox River onto the Wisconsin River and then the Mississippi River. Paddling back against the current, they were taught an easier return route by local Native Americans up the more placid Illinois and Des Plaines rivers and onto the Chicago River. A portage crossed the low continental divide separating the Des Plaines River and the Chicago River, the latter of which emptied into Lake Michigan. From here they paddled up Lake Michigan to St. Ignace, where Jolliet carried on without Marquette to Montreal.

Having personally traveled most of the twenty-three-hundred-mile route from the St. Lawrence River to the Great Lakes and the fifteen-hundred-mile Mississippi River route connecting the Gulf of Mexico to the Chicago portage, Jolliet saw great value in a continuous water route and proposed a short canal

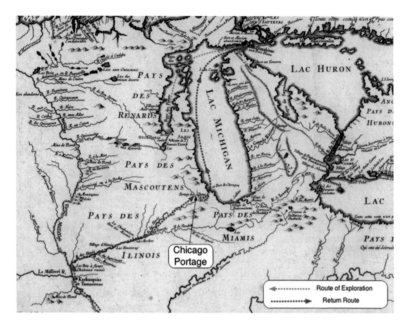

Marquette and Jolliet's route of exploration to the Mississippi River via Green Bay and return route through the Chicago portage. © Patrick McBriarty

low body of water called Portage Lake, or Mud Lake. It varied in depth and length according to the wetness of the season and stretched five to six miles in length, running parallel to the portage road. Occasionally, the Des Plaines would flood its banks to connect with Mud Lake and in turn overflow into the Chicago River, creating a continuous water route navigable by canoe.

The year after Marquette and Jolliet first passed through Chicago, development of the fur trade in the area began. Explorer René-Robert Cavelier, Sieur de La Salle, utilized the portage several times, as did his famous right-hand man, Henri de Tonti. From the 1680s through the 1690s, at least one hundred French traders, explorers, and missionaries passed through or settled in the region to establish various depots, forts, and the Mission de l'Ange Gardien (Mission of the Guardian Angel) along the Chicago River.

Throughout most of the eighteenth century, growing hostilities, including two protracted wars between the Fox Indians and the French (the First Fox War [1712–16] and Second Fox War [1728–33]) and the French and Indian War (1754–63), destroyed all settlement and trade in Chicago. This route to the West had national importance for the young United States; however, Chicago and its portage remained under the control of hostile tribes until after the Treaty of Greenville and the establishment of Fort Dearborn in 1803, initiating U.S. control of the area. In 1795 the Treaty of Greenville secured from the Native Americans a six-square-mile piece of land at the mouth of the Chicago River and free passage of the portage and waterways to the Mississippi River. The first Fort Dearborn was destroyed at the beginning of the War of 1812 but was reestablished in 1816 as settlers returned.

As described by Chicago fur trader Gurdon Hubbard in 1818, navigating the portage could be quite an ordeal. Upon reaching

to replace the Chicago portage. Chicago's location at a natural break in transportation—combined with continuing westward settlement and rapid development of shipping and trade—would ensure that the Chicago River became a busy waterway and nurture a great city.

The "Chicagou" Portage (as it was labeled on early French maps) was a key link between waterways. Its eastern end began at the West Fork of the South Branch of the Chicago River near the modern-day intersection of 26th Street and Western Avenue. The seven-mile portage road ran southwest over the low, marshy ground and ended at an eastward bend in the Des Plaines River, near today's intersection of Harlem Avenue and 49th Street. Often used by Native Americans, and then increasingly by western Europeans after 1673, the portage also included the long, shal-

the portage, boats were dragged through the shallow channel or, when the channel was dry, placed on short rollers and shoved or pulled up to a mile and a half to Mud Lake. Once there, four men would get in and pole each boat,

> while six or eight others waded in the mud alongside, and by united efforts constantly jerking it along. . . . While part of the crew were thus employed, others busied themselves in transporting our goods on their backs to the river. . . . Those who waded through the mud frequently sank to their waist, and at times were forced to cling to the side of the boat to prevent going over their heads; after reaching the end and camping for the night came the task of ridding themselves of the blood suckers [leeches]. . . . Having rid ourselves of the blood suckers, we were assailed by myriads of mosquitoes, that rendered sleep hopeless. . . . It took us three days of toil to pass all our boats through this miserable lake.[1]

Plans for a canal 150 years after it was first proposed by Jolliet resulted in real-estate speculation and city boosterism that spurred the early development of Chicago. In 1822 Congress made the first appropriation granting land for the construction of the Illinois & Michigan (I&M) Canal to replace the Chicago portage. In 1825 the opening of the Erie Canal, connecting the Hudson River and New York to the Great Lakes, also encouraged the development of Chicago. This led to eastern financing and New York connections, as Chicago was seen as the next leaping-off point for westward expansion. Not surprisingly, the City of Chicago's first mayor, William B. Ogden, arrived from New York in 1835 to oversee family real-estate investments in Chicago. By the mid-1830s, most Native Americans had been forced west of the Mississippi, essentially ending Chicago's early French-Indian culture.

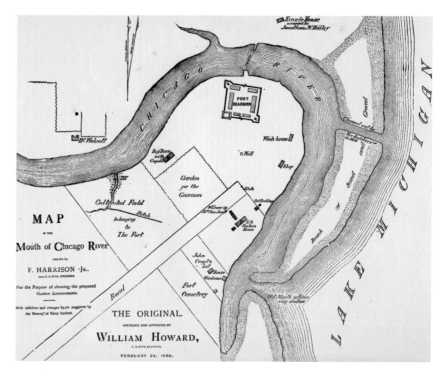

Map of the entrance and mouth of the Chicago River in 1830.

During this time, the soldiers at Fort Dearborn, who often received supplies by ship, made several efforts to free the mouth of the Chicago River from the shifting sand dunes on the lakefront. In 1834 army engineers permanently opened the river to create Chicago's harbor. That year 176 ships arrived, with 250 vessels the next and 1,427 sailing ships and 39 steamships the following year. This established the Chicago River as an important harbor and waterway.

The permanent harbor, growing ship traffic, and the increase in trade and population meant greater demand for improved transportation; not surprisingly, Gurdon Hubbard was a big proponent of building the Illinois & Michigan Canal, which finally opened in 1848 to replace the portage. This new water route connected through fifteen locks from the West Fork of the South Branch to the Illinois River at LaSalle, Illinois. A complete waterway now linked the eastern cities of New York, Boston, Philadelphia, and Baltimore to St. Louis, New Orleans, and points in between. In Chicago, and specifically in Bridgeport at the eastern end of the I&M Canal, freight was transferred from ship to canal barge and vice versa on the way to its final destination. The canal was deepened in 1871 to accommodate larger barges and improve the flow of water and sewage away from Chicago. These improvements created a canal that was sixty feet wide and six feet deep for the entire ninety-six-mile length. This secured the Chicago River's position for the next hundred years as a key waterway to the city and the nation.

The I&M Canal increased western trade and more reliably brought agricultural goods to Chicago. Imagine a farmer driving his wagon team, hoping to be first to market, gambling that rain would not turn the roads to mud and strand him short of town while his year's harvest rotted. A canal barge leading straight to Chicago that saved time and worry would be a great advantage to the agricultural production of the region. Suddenly, there was a reliable link to Chicago markets and the waiting eastern trade goods that arrived via Great Lakes ships. The improvement in transportation accelerated growth. The continuous waterway through Chicago linking the Great Lakes and the Gulf of Mexico dramatically bolstered Chicago's prospects as the gateway to the West. The I&M Canal solidified the importance of the Chicago River as a port and waterway and reinforced shipping rights. Although the canal was officially closed in 1933 and then filled in within the city limits, the City of Chicago maintained right-of-way along this corridor. This same strip of land now carries the Stevenson Expressway (I-55).

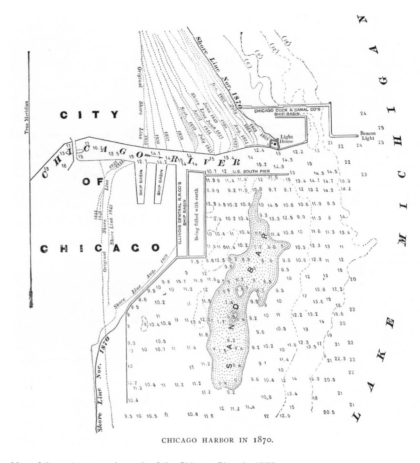

CHICAGO HARBOR IN 1870.

Map of the entrance and mouth of the Chicago River in 1870.

Long since established by canoe and now canal boat, the transportation route on this relatively narrow, low-banked waterway through the prairie landscape effectively prohibited fixed bridges on the Chicago River except along its farthest reaches. With shipping growing by leaps and bounds, a typical fixed bridge would cripple Chicago's future. Construction of a high fixed bridge with a 100- or 150-foot clearance for sailing ships was far too costly in time, materials, and land; such a fixed bridge would require long approaches to carry horse-and-wagon teams

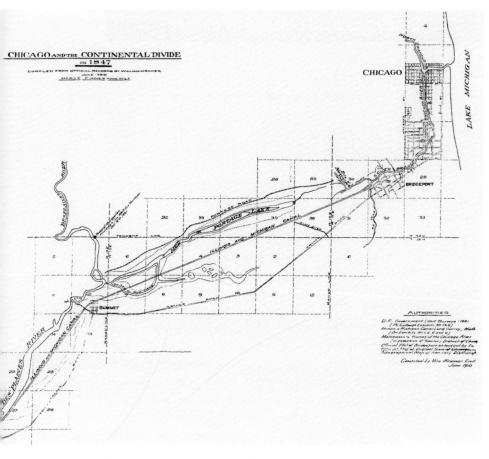

Map of the Chicago portage and Illinois & Michigan Canal in 1848.

pany to tunnel under the river. However, a new ordinance requiring tugboats to use hinged smokestacks in order to pass under the bridges afforded enough relief that the company was later dissolved. The idea was revisited a decade later, and several tunnel proposals were presented to the city council between 1864 and 1866.

In 1868 the city council approved the first tunnel at Washington Street. Two additional tunnels followed, at La Salle Street in 1871 and Van Buren Street in 1892. The drivers of heavily loaded wagons disliked using the tunnels, however, because horse teams had to slow or stop to rest partway up the ramped roadways. There was also much less light carriage and pedestrian traffic than expected, as the tunnels were found to be dark and murky. At first a seemingly superior alternative uninterrupted by ship traffic, the tunnels never really offered significant competition to city drawbridges. These tunnels would, however, become important to the streetcar companies, and new tunnels for the city subway system would also be constructed under the river decades later.

The geology and geography of the area underpinned Chicago's need for drawbridges. Its slow-moving river and location on the glacial plains at the southwest corner of the Great Lakes made it an ideal jumping-off point for westward travel and expansion. The short seven-mile Chicago portage over the low continental divide and then the I&M Canal that connected to the Chicago River just five miles upstream from Lake Michigan made Chicago important to both the Eastern Seaboard and the agricultural West. Named "Nature's Metropolis" by William Cronon, despite the wet, swampy ground, Chicago was ideally situated along a natural transportation route at a key transfer point and became a center for commerce, industry, and urban-

up and over it, which, even in the best weather, would prove problematic. On the narrow Chicago River, drawbridges made the most sense. These less intrusive and much more cost-effective moveable bridges enabled both ship and street traffic (though not at the same time) to ensure Chicago's continued growth.

Yet, with increasing frequency, open drawbridges prompted consideration of other means of crossing the river. Tunnels were first proposed in the early 1850s, and William Ogden, William Gooding, Edward Tracy, and Thomas C. Clarke formed a com-

ization. Traders, merchants, and then industry expanded along the river and depended on the waterway for the shipment of goods and supplies. The early grain, lumber, and meatpacking houses of Chicago demanded waterfront property. Because the Y-shaped river naturally divided Chicago into North, West, and South sides, it was the drawbridges that helped to create and connect the city's disparate parts into one developing whole. There was no reasonable alternative to moveable bridges.

## Why Chicago Built So Many Drawbridges

Chicago's bridge history began in 1832 when the first bridge was built, more than a year before Chicago was incorporated as a town. A second bridge was constructed in the winter of 1832–33 to carry wagon teams across the South Branch. In 1834 Chicago's third bridge—and first drawbridge—was built across the river's main channel at Dearborn Street. Crossing the waterways was essential, and the Chicago bridges had to meet the changing demands of the city. They were adapted from supporting pedestrian and horse-drawn vehicles to also accommodating railroad and streetcar traffic and, eventually, truck and automobile traffic, all the while opening for ships. In less than two hundred years, more than 350 bridges were built to maintain the 150 crossings over the forty-two miles of Chicago waterways. At least 200 of these were drawbridges. Chicago quickly became the drawbridge capital of the world.

Growth of the city required new and better river crossings and replacement of worn-out or inadequate bridges. Chicago, centered on the South Side, quickly expanded along the northern and southern branches of the Chicago River, leading to exponential growth. With heavy use, wood bridges soon gave way

| Development of Chicago by decade | | | |
|---|---|---|---|
| City of Chicago | | Chicago River bridges | |
| Decade | Population | Draw | Fixed |
| 1830 | 1,000* | 0 | 0 |
| 1840 | 4,853 | 3 | 0 |
| 1850 | 29,963 | 7 | 2 |
| 1860 | 109,260 | 16 | 1 |
| 1870 | 298,977 | 26 | 0 |
| 1880 | 503,185 | 36 | 1 |
| 1890 | 1,099,850 | 48 | 4 |
| 1900 | 1,698,575 | 57 | 11 |
| 1910 | 2,185,283 | 60 | 13 |
| 1920 | 2,701,705 | 60 | 15 |
| 1930 | 3,376,438 | 68 | 18 |
| 1940 | 3,396,808 | 71 | 17 |
| 1950 | 3,620,962 | 72 | 16 |
| 1960 | 3,550,404 | 73 | 18 |
| 1970 | 3,369,359 | 67 | 24 |
| 1980 | 3,005,072 | 61 | 28 |
| 1990 | 2,783,726 | 61 | 29 |
| 2000 | 2,896,016 | 57[†] | 31 |
| 2010 | 2,695,598 | 55[†] | 32 |

Table includes both City and railroad drawbridges.

*Estimated to be approximately 80 percent Native Americans.

[†]Several of these drawbridges are now fixed in place and no longer open.

under the increasing traffic that necessitated their replacement with more durable structures and designs.

The city limits of Chicago began expanding in 1847 to incorporate neighboring lands and nearby towns. The most dramatic additions came in 1889 with the annexation of the Lake View,

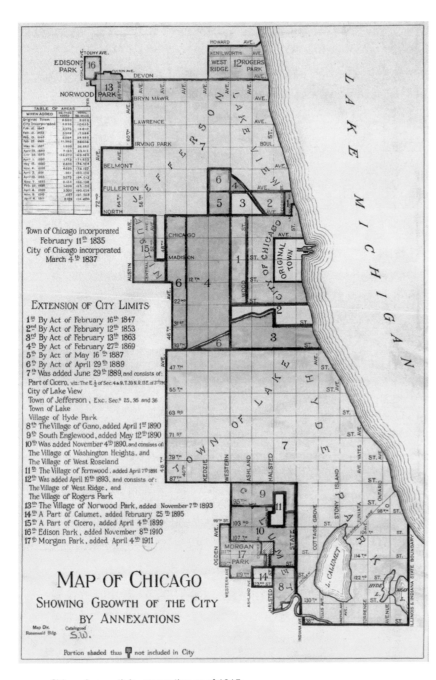

Town of Chicago incorporated
February 11th 1835
City of Chicago incorporated
March 4th 1837

EXTENSION OF CITY LIMITS

1st By Act of February 16th 1847
2nd By Act of February 12th 1853
3rd By Act of February 13th 1863
4th By Act of February 27th 1869
5th By Act of May 16th 1887
6th By Act of April 29th 1889
7th Was added June 29th 1889, and consists of:
Part of Cicero, viz.: The E. ½ of Sec. 4 & 9, T. 39 N.R. 13 E. of 3rd P.M.
City of Lake View
Town of Jefferson, Exc. Sec.s 25, 35 and 36
Town of Lake
Village of Hyde Park
8th The Village of Gano, added April 1st 1890
9th South Englewood, added May 12th 1890
10th Was added November 4th 1890, and consists of:
The Village of Washington Heights, and
The Village of West Roseland
11th The Village of Fernwood, added April 7th 1891
12th Was added April 19th 1893, and consists of:
The Village of West Ridge, and
The Village of Rogers Park
13th The Village of Norwood Park, added November 7th 1893
14th A Part of Calumet, added February 25th 1895
15th A Part of Cicero, added April 4th 1899
16th Edison Park, added November 8th 1910
17th Morgan Park, added April 4th 1911

MAP OF CHICAGO

SHOWING GROWTH OF THE CITY
BY ANNEXATIONS

Map Div.
Rosenwald Bldg.     Catalogued
S.W.

Portion shaded thus ▓ not included in City

Chicago's growth by annexation as of 1915.

Jefferson, Hyde Park, and Lake townships. This additional land included the Calumet River and Harbor and a significant stretch of the North Branch of the Chicago River, further increasing the City's bridge inventory and responsibilities.

Meanwhile, Chicago's first railroad, the Galena & Chicago Union Railway, was begun in 1848, and in 1852 the first railroad bridge crossing the river was built near Kinzie Street. Over the next few decades, development of the railroads made Chicago the most important railroad hub in North America. This necessitated additional drawbridges over the Chicago River and other city waterways. The eastern and western railroads met in Chicago, furthering its status as a center for trade, commerce, and development. Additionally, the growth of the railroads spurred technological advancements in material science, engineering, and construction, and a key element of their expansion was bridge innovation and design.

Repeated efforts to deepen and widen the river also forced additional bridge building. Between the 1860s and 1870s, the South Branch was dredged, lengthened, and widened, and collateral slips and canals, including a mile-long canal serving the Union Stock Yards, were added. In an age totally lacking environmental protection laws, the unchecked development soon turned the Chicago River into little better than an open sewer. By 1889 health concerns initiated the charter of the Chicago Sanitary District (now the Metropolitan Water Reclamation District of Greater Chicago) to address water sanitation. The resultant Chicago Drainage Canal, later renamed the Sanitary & Ship Canal, would replace the I&M Canal.

The Sanitary District, an independent public agency created by the State of Illinois, undertook the extraordinary engineering feat of reversing the flow of the Chicago River. The city had always drawn its supply of freshwater from Lake Michigan, and reversal of the river would help protect the lake from waste and runoff. A massive excavation and engineering project, the Sanitary District reversed the river so it flowed away from Lake Michigan in 1900. Since then, water from Lake Michigan has flowed into the Main Channel and into the South Branch and via the Sanitary & Ship Canal into the Illinois and Mississippi rivers. The project also widened much of the Chicago River to a width of two hundred feet to accommodate larger ships. The "Lakes to Gulf" route opened in July 1910, connecting the Great Lakes and the Gulf of Mexico for ships and barges. Somewhat smaller in scope, the eight-mile North Shore Sanitary Canal began diversion of the waste of the northern suburbs into the North Branch of the Chicago River that same year. The Sanitary District constructed the canal from Lake Michigan in Wilmette to connect with the North Branch just east of Foster and Kedzie avenues. This led to extensive construction of fixed bridges by the Sanitary District across this new canal in the city and suburbs.

Wholesale bridge replacement was accelerated three times during Chicago history. The flood of 1849 destroyed all five of Chicago's bridges, and the Great Chicago Fire of 1871 destroyed eight downtown bridges. Finally, navigational interests supported by federal oversight required replacement of most of the city's swing bridges between 1890 and 1930.

## How Chicago Paid for Its Bridges

Throughout history it has been in Chicago's best interests to build new and better bridges, yet bridges cost money. Despite the obvious benefits of spurring business, industry, and real-estate development, bridges involve a variety of costs, from plan-

ning, design, construction, and land acquisition to maintenance, operation, and repair.

Before Chicago had a local government, its first two bridges  were built by residents. Tavern owner Samuel Miller built the first bridge; another tavern keeper, Charles Taylor, and his brother Anson built the second after raising subscriptions from locals. On August 10, 1833, Chicago incorporated as a town, and bridge maintenance and oversight fell to the newly elected Common Council (the precursor to today's Chicago City Council). That December a Bridge Committee was created to advise and oversee repair of the town's two bridges. The town's third bridge at Dearborn Street, built in 1834, was also privately funded. Chicago's incorporation as a city on March 4, 1837, continued local government responsibility for bridge maintenance, while bridge construction remained almost entirely privately financed. This was consistent with the overriding principle of government during the nineteenth century that the cost of public projects should be borne by the users.

In 1847 the Common Council appointed a special committee to oversee the construction of the first bridge at Madison Street. The Common Council used special bridge committees again after the flood of 1849 to oversee rebuilding of the town's five bridges. An additional special committee was appointed to borrow funds for bridge construction. It was during this crisis that the first municipal funds, amounting to fifteen hundred dollars, were appropriated toward construction of the second Madison Street Bridge. However, residents, businessmen, and landowners shouldered the bulk of the bridge replacement costs. Going forward, local bridge committees would organize to gather private and political support and manage subscription lists for new crossings or bridge replacement. Pledged contributions for a bridge were gathered and recorded by simply denoting the contributor's name and pledged amount, thus creating the bridge subscription list.

Chicago also reused bridge components and complete bridges whenever possible. In 1849 the boiler iron floats of the old Madison and Wells Street bridges were reused on new bridges. In 1854 the first Polk Street Bridge was built using the old Clark Street Bridge. Similarly, in 1856 the old Randolph Street Bridge was moved to become the first North Avenue Bridge, and a new Randolph Street Bridge was built in its place. Occasionally, a busy crossing with an inadequate bridge received a longer, wider span. On these occasions, if still serviceable, the old bridge was moved and reused, usually to create a new crossing in a growing but less trafficked area.

After receiving City approval, the railroad companies financed and built their own bridges. Occasionally, the City was able to convince railroads to contribute funds for highway bridges when it was in both parties' best interests. The first example of this was the construction of the Rush Street Bridge in 1856. The Illinois Central (IC) Railroad, founded in 1851, opened its line from Chicago to New Orleans in 1856, which terminated at the lakefront south of the river by Michigan Avenue. Meanwhile, the new Galena & Chicago Union (G&CU) Railroad terminal was on the north side of the river. This put the Chicago River between the two railroads. The IC and G&CU railroads and the City of Chicago contributed equally to the first Rush Street Bridge, a critical connection for passengers and freight. Similarly, the current Monroe Street Bridge, completed in 1919, was also funded by the railroads in conjunction with the building of Chicago's Union Station.

By the mid-1850s, the City began to provide up to half the cost of each bridge, and private subscriptions covered the re-mainder. In 1857 the third Madison Street Bridge was controversial as the first bridge entirely financed by public funds. The practice of subscription lists and private financing did not simply disappear, however. Bridge construction and maintenance were still expensive propositions. The City adopted a policy whereby private financing was expected to pay for at least half the cost of the first two bridges at any river crossing; thereafter, the City would pay for construction and maintenance of additional bridges. For example, in 1862 the cost of the second Clybourn Place Bridge (now Courtland Street) was built with private subscriptions and City funds, whereas the City paid for the entire cost of the third Courtland Street Bridge in 1873.

In 1862 payment issues between the contractor and subscribers for the first South Halsted Street Bridge led to subscription lists, once final, being handed over to the City. Thereafter, the city clerk and treasurer extended funds, hired and paid contractors, and collected funds from subscribing residents. Over the following decades, the process of special property-tax assessments evolved to replace the local bridge committees and subscription lists. This soon allowed more sophisticated financing practices such as the issuance of bridge bonds backed by special property-tax assessments. Thus, voluntary contributions ended, and the City taxed the expected benefactors of a new bridge, the surrounding property owners, via special assessments. It was understood that a bridge would compensate taxpayers by way of increased business, higher rents, or both.

In 1870 a new Illinois State Constitution was adopted that allowed municipalities to assume indebtedness up to 5 percent of annual assessed property values. From then on, the City assumed all costs for river-bridge construction; whenever possible, though, the City leaned on railroads and streetcar companies to contribute funds for highway bridges. For the first time, the City of Chicago was able to take on long-term debt to finance public works. Prior to this, city government was run on a "pay as you go" basis, and bridge construction, maintenance, and operating expenses had to be funded out of the annual budget.

Surrounding municipalities also sometimes paid for Chicago bridges. In 1874 the City shared financing for the first bridge at Fullerton Avenue with the town of Lake View to the north, as Fullerton Avenue was the border between two municipalities. Also, many bridges built outside Chicago limits would later be incorporated, such as the first Belmont Avenue Bridge. Built in 1875, it was funded by Cook County and the towns of Lake View and Jefferson. Added to Chicago by annexation in 1889, Belmont Avenue and other bridges would add to the City's bridge inventory and maintenance costs.

By 1880 Chicago had thirty-three bridges over the Chicago River and its branches. All were center-pier swing bridges. Seven of these bridges, at State, Clark, Wells, Randolph, Madison, Van Buren, and South Halsted streets, carried street railways, and an eighth at Archer Avenue carried both a street railway line and the Chicago & Alton Railroad. No bridge was built that year, and the entire maintenance budget for bridges and viaducts was $61,847.85. In relative terms, these funds would pay for half of the two-lane, all-steel Rush Street Bridge built in 1884, or for six wood and iron bridges such as the 1877 North Halsted Street Bridge. Today, about $10 million is spent annually to maintain and operate city bridges, and construction of a new fixed bridge costs more than twice that amount.

By 1890 the City of Chicago was responsible for forty-four swing and eight fixed bridges over the I&M Canal and the Calu-

met and Chicago rivers. Soon thereafter, the City and the Sanitary District would begin replacing the obstructive swing bridges. The Sanitary District launched a $2.5 million bond issue in 1900 to fund "river improvements, including erection of bascule bridges in place of center pier structures."[2] The Sanitary District also constructed and paid for railroad and highway bridges crossing the new Sanitary & Ship Canal between 1898 and 1910. The Sanitary District, funded through property taxes, built an additional eleven bridges, in agreement with the City, over the Chicago River between 1902 and 1907.

An Illinois Supreme Court ruling in February 1901 limited the City of Chicago's municipal debt, and only $1.25 million could be secured for the city's first bascule bridges. In 1904 the City was able to issue an additional $5.5 million in bonds to fund public works projects, which included construction of more bridges. By 1910 ten Chicago-type bascules had been completed. The following year, the City received voter approval to issue $4,655,000 in new bridge bonds, and between 1913 and 1922 thirteen additional bridges were completed.

In 1937 the City negotiated appropriations from the Motor Fuel Tax Fund, initiated by the State of Illinois in 1927. The City successfully argued that the law's articles applying to the maintenance of bridges on arterial streets included Chicago bridges, and twenty-five of the City's eighty-two bridges were so located. That October $142,000 in state funds was obtained, initiating an annual appropriation from the state Motor Fuel Tax Fund to the City of Chicago for the local bridge and maintenance fund.

Chicago also took advantage of federal funds whenever available. The Lake Shore Drive Bridge, completed in 1937, was part of a huge lakefront development project undertaken by the Chicago Park District and funded by the Public Works Administration. The PWA was part of President Franklin Delano Roosevelt's New Deal program to help move the country out of the Great Depression and create jobs. The South Ashland Avenue Bridge, built in 1938, also used PWA funds, and the State Street Bridge was constructed with a 45 percent PWA grant and completed in 1949.

Funding for about half of the subsequent Chicago River bridges came either from the City's share of the Motor Fuel Tax Fund or from state and federal sources. The North Ashland Avenue Bridge, built in 1936, used a combination of these and PWA moneys, whereas the 1938 Torrence Avenue Bridge and the 1940 South Western Avenue Bridge were funded primarily by the City's share of the Motor Fuel Tax Fund. Construction of the Ohio Street and Congress Parkway bridges in the 1950s and '60s was paid for entirely by state and federal funds, as they were both part of the interstate highway system (both are owned and maintained by the State of Illinois, but operated by the City of Chicago). Similarly, the last new drawbridge at Randolph Street was built with $18 million in state and federal funds in 1984, but was designed, owned, and operated by the City of Chicago.

The $21.4 million fixed bridge on North Avenue was aided by a $5 million appropriation from the 2004 Transportation Appropriations Bill, allocated through the Federal Highway Administration's Highway Bridge Replacement and Rehabilitation Program. The balance of the funds came from the City of Chicago. Although the City is responsible for operating and maintaining Chicago's highway bridges, state or federal funds are generally required to fund new bridge construction because of rising replacement costs.

## The Rise and Fall of Chicago Drawbridges

Drawbridge construction peaked in the 1870s and rebounded again in the first decade of the twentieth century. The growth of the city and the short life spans of all-wood and wood and iron bridges drove this; replacement of the swing bridges created a second spike in drawbridge construction. New technology, materials, and better designs increased the life span of drawbridges from four to five years in the 1840s to forty to fifty years in the 1880s. By 1900 Chicago had fifty-seven working drawbridges, nearly fifteen more than it does today. The number of working drawbridges would continue to rise and peaked at seventy-three by 1960.

After the establishment of the Board of Public Works in 1861, the City Bridge Department began managing bridge planning, construction, and maintenance. A major problem during this time was damage caused to bridges by commercial ships. Collisions caused a great deal of repairs and were all too common due to human error. The City and the shipping companies sued each other for damages whenever fault could be proven on the part of either the bridge tender or the ship's captain. This wear and tear on the swing bridges encouraged new bridge building.

Although drawbridge development and construction was at an all-time high, ship traffic was about to enter a period of decline. The nature of shipping on the Great Lakes was changing; although more than 21,000 ships arrived and departed from the Chicago River in 1887, a trend toward increasingly larger steel ships that could carry greater cargo loads was under way. This reduced the number of ships on the Great Lakes and the Chicago River. Although total tonnage continued to increase until 1893, the number of ships on the river began a steep decline, and

## Bridges built over the Chicago River by decade

| Decade | Moveable bridges built | Fixed bridges built |
|--------|------------------------|---------------------|
| 1830s | 3 | 6 |
| 1840s | 11 | 5 |
| 1850s | 16 | 1 |
| 1860s | 21 | 1 |
| 1870s | 28 | 5 |
| 1880s | 14 | 3 |
| 1890s | 14 | 11 |
| 1900s | 21 | 3 |
| 1910s | 11 | 11 |
| 1920s | 8 | 5 |
| 1930s | 10 | 5 |
| 1940s | 2 | 0 |
| 1950s | 3 | 0 |
| 1960s | 4 | 7 |
| 1970s | 1 | 3 |
| 1980s | 2 | 1 |
| 1990s | 0 | 1 |
| 2000s | 0 | 1 |
| 2010s | 0 | 1 |
| Total | 169 | 70 |

by 1900 just 9,788 ships came in and out of the river; by 1920 only 2,545 ships, carrying just 16 percent of the total tonnage of 1887, used the river.

Government officials reacted slowly, failing to construct better drawbridges to create a wider river channel and allow the largest Great Lakes ships. This constrained the development and location of Chicago industry. Urban infrastructure, including bridges, was not meeting the needs of the city and the changes in shipping, which encouraged a move by heavy industries away from the Chicago River in favor of the Calumet River and Harbor. Developed after the Great Fire in 1871, the Calumet River soon competed for Chicago River shipping and offered large tracts of open land for commercial development. Lower real-estate pricing in the developing Calumet area proved attractive to Chicago's growing heavy industries and big business. In 1906 the Calumet River surpassed the Chicago River in total tonnage and by 1916 exceeded its number of ship arrivals and departures.

After 1911 federal authorities required a clearance of sixteen and a half feet on all new drawbridges over the Chicago River. With this standard in place, the remaining commercial traffic on the river transitioned to barges and tugboats that could pass freely under the bridges. Most Chicago tugboats now have a hydraulic lift under the wheelhouse to raise them for better visibility and lower them to pass under the bridges.

Yet substantial drawbridge construction continued into the 1920s and '30s and was considered a greatly needed element of the city's modern infrastructure. In 1921 the Calumet Harbor gained federal approval to become a deepwater port, led by longtime area investor and booster James H. Bowen. That year more than 80 percent of Chicago's shipping tonnage and 73 percent of incoming ships used the Calumet River and Harbor. Despite a brief uptick in shipping during World War II, however, commercial traffic on the Chicago River has been on the wane since 1921.

The City Bridge Department was responsible not only for operating and maintaining all highway bridges over the waterways, but also for viaducts and street overpasses. Thus, the cost advantage of a fixed bridge compared to a drawbridge was naggingly apparent throughout Chicago bridge history. As early as 1880, City officials proposed a fixed-bridge policy and the development of an extensive lakefront harbor as an alternative to using the Chicago River as a harbor. This plan would have allowed the downtown drawbridges to become fixed. Though often revisited, the plan was never implemented, and the lakefront was preserved for public use. Although Municipal Pier (now Navy Pier), completed in 1916, was primarily added to the lakefront in support of shipping, it was also available for public use as an escape from the city and the enjoyment of the cool lake breezes in an era before air-conditioning. Used during World War II as a naval training center, afterward it was returned to its original use for commercial shipping and freight transfer, which ended in the late 1960s. Employed for a variety of uses since, it fell into disrepair. In the 1990s, it underwent a two-hundred-million-dollar remodeling and is now Chicago's number-one tourist attraction.

In the early 1900s, the City began promoting a fixed-bridge policy on the North Branch. Chasing the promise of significant reductions in bridge maintenance and operating costs, the City embarked upon a sixty-year effort to gain federal approval of the policy. In-depth surveys and impact studies were presented in 1924 and 1925, but the requests for federal approval were consistently denied.

A notable U.S. Navy contractor and ship builder, Henry C. Grebe & Company, was located on the river just north of Belmont Avenue. The existence of this shipyard on the North Branch was likely a significant reason for the continued need for moveable bridges on the North Branch, up to and including the Belmont Avenue Bridge. The existence of this shipyard benefited the nation during World War II, as Great Lakes shipbuilding made a significant contribution to the war effort. Grebe & Company

would produce twenty-five minesweepers, nineteen tugs, and three water tankers for the U.S. Navy during the war. They would also construct three additional minesweepers in 1953–54 for the Korean War and one prototype landing craft swimmer reconnaissance ship in 1968. Navy underwater demolition teams, also know as SEAL teams, used this prototype boat for insertion and retrieval during the Vietnam War.

After World War II, shipping on the Chicago River, particularly the North Branch, declined significantly. In May 1950, the Chicago City Council passed a resolution in favor of immobilizing the eight existing moveable bridges above North Avenue, but received no response from federal authorities. Three years later, the City compiled an exhaustive survey and again proposed a fixed-bridge policy on the North Branch. It reported that the only large vessels requiring bridge openings north of North Avenue were minesweepers, and the last contracted Grebe minesweeper was to be delivered that same year. Most of the bridge openings were for recreational sailboats and construction barges, and the City argued that it would be no great hardship for this traffic to run under a twenty-one-foot fixed bridge. For example, in 1954, the Belmont Avenue Bridge opened only 83 times, compared with 1,369 openings in 1907. And, as the report stated, "insofar as the mine-sweepers are concerned, the shipyard can be relocated elsewhere in the Chicago area if it continues to be of importance to the National Defense program."[3]

City arguments finally prevailed in 1955, and the Army Corps of Engineers agreed that the bridges above North Avenue were essentially operating as fixed spans. The greatest decline in drawbridge use was on the North Branch, where the loss of ship traffic left many of the bridges sitting years between openings. In 1961 the first North Branch drawbridge was replaced by a fixed span at Fullerton Avenue. A similar proposal was made for a controversial fixed bridge over the South Branch during the 1950s that would act as part of the interstate highway system. Although meant to provide a sixty-foot river clearance, the plan generated vigorous protests from shipping interests for several years before finally being approved in 1961. The high fixed bridge was constructed as part of the Southwest Expressway, later renamed the Dan Ryan Expressway, and opened in 1965. No other fixed bridges have been built over the navigable portion of the South Branch since.

Following the 1970s, the total number of operational drawbridges began to gradually decline. At the beginning of the decade, three moveable bridges at Throop Street, Polk Street crossing the South Branch, and Erie Street crossing the North Branch were removed and not replaced. Eight North Branch drawbridges were replaced by fixed spans between 1973 and 2012. The removal of the Ogden Avenue Overpass in 1994 also resulted in the removal of two additional Chicago-type bascule bridges without replacement.

The fixed-bridge policy on the North Branch was expanded in the 1990s, and at first the City was permitted to treat the remaining drawbridges above North Halsted Street and on the east side of Goose Island as fixed spans. Several years later all City drawbridges on the North Branch were converted to fixed spans. By bolting I-beams across the center of these structures, the leaves were permanently locked together. Operating machinery was either removed or abandoned, with two conversions including dismantling of the then unnecessary bridge houses.

The City has also added restrictions to navigation on the Chicago River and instituted various cost-saving measures, such as the roving bridge-tender crews and conversion of bridges to one-person operations. What started in the 1970s and 1980s as informal cooperation between the boatyards and the bridge operators was formalized in 1995, and the revised Code of Federal Regulations concerning the Chicago River now requires advance notification for bridge lifts. The biannual river trips made by recreational boats to and from Lake Michigan are now scheduled and require the City to open the bridges to river traffic only two to three days per week in the spring and fall.

## Chicago's Contribution to Moveable-Bridge Development

By the 1880s, Chicago featured the greatest concentration of drawbridges in the world. Though initially constructed by master carpenters and masons, most of Chicago's bridge builders were practical men, self-taught or trained through trade-apprenticeship systems passing down the skills and art of bridge building. Such men would build most of the city's street, highway, and railroad bridges well into the late 1800s.

The United States Military Academy at West Point was the first school in the United States to teach engineering, beginning in 1824. The origin of bridge engineering in the United States is often marked in 1847 after Squire Whipple, an engineer from upstate New York, self-published *A Work on Bridge Building*. This book provided the first accurate mathematical stress analysis of a bridge truss. Formal training in engineering was accelerated by the American Civil War. As the engineering profession grew, it spilled over into bridge building and design. Several decades after the Civil War, engineers would gain primacy over and standardize bridge building and design. The best and brightest builders, engineers, architects, and inventors were attracted by the rapid growth of the industry that contributed so much to the vibrancy of Chicago. Not surprisingly, Chicago developed a strong bridge-building industry and became a center for bridge engineering.

John Van Osdel, considered Chicago's first architect, designed and built an early pontoon swing bridge in 1841 at Wells Street. During the 1850s, Derastus Harper, superintendent of public works for the City of Chicago, designed and constructed several early wood pontoon swing bridges and the very first Chicago pivot bridges at Lake, Clark, and Wells streets. A decade later, John K. Thompson, a city commissioner of public works, provided important swing-bridge improvements. Another important self-taught bridge builder in Chicago was William Howard of the firm Fox & Howard, a company that dominated bridge building in Chicago from 1860 to 1872. Between the 1850s and the early twentieth century, four or five firms—two or three of them often Chicago based—would bid usually on any given bridge project, highlighting the competitiveness of the profession nationwide. Similarly, construction of America's rail and road infrastructure during this time involved many Chicago-based bridge builders and engineers.

William Scherzer, a Chicagoan and Swiss-trained engineer working in the latter half of the nineteenth century, patented one of the very first modern bascule designs. Tragically, he died at the age of thirty-five and received his patent posthumously. His brother Albert, however, originally an attorney, would use William's design to create the Scherzer Rolling Lift Bridge Company. The Chicago-based company developed fourteen additional bridge patents and built more than 175 Scherzer bridges nationwide by 1916.

Joseph B. Strauss moved to Chicago after attending the University of Cincinnati and developed more than thirty bridge patents throughout the early twentieth century, including several bascule-bridge patents. Characterized as egotistical and self-serving, Strauss spent several years and a great deal of his own money to promote the building of a bridge across San Francisco Bay and is best known as the chief engineer of the Golden Gate Bridge.

Swedish-born John Ericson also made his mark on Chicago. Ericson, supported by talented City bridge engineers like Alexander von Babo and Thomas Pihlfeldt, developed the Chicago-type bascule design. Over their careers, Pihlfeldt and von Babo combined to provide more than seventy-seven years of design and engineering service to the City of Chicago. Other notable bridge inventors and engineers also working in Chicago included Captain William Harman, Ralph Modjeski, Theodor Rall, John Page, Hugh Young, Donald Becker, and Stephen Michuda.

Because of this abundance of talent, many new innovations, technologies, and materials were attempted first in Chicago. In regard to moveable bridges, these include:

- The first turntable pontoon swing bridge, completed in 1849
- The first all-iron swing bridge west of the Alleghenies, opened in 1856
- The first large-scale vertical-lift bridge, completed in 1894
- The first Scherzer rolling-lift bridge, designed in 1893 and built in 1895
- The first Chicago-type bascule bridge, opened in 1902
- The longest single-leaf bascule bridges, successively built in 1908, 1913, and 1919
- The first double-deck, double-leaf drawbridge, completed in 1916

- The world's largest bascule bridge when completed in 1937
- The world's second-largest bascule bridge, completed in 1982

Chicago's advancement of bridge technology and position at the cutting edge of bascule-bridge design is also evidenced in the number of U.S. bridge patents with Chicago origins. Chicagoans have contributed more than 100 of the approximately 350 patents on drawbridges. Within those, Chicagoans generated nearly 60 percent of the 120 bascule-bridge patents, by far the highest percentage from any city in the world.

## The History of Chicago Drawbridges
### Chicago River Ferries

For centuries a canoe was the usual means to reach and cross the Chicago River. Antoine Ouilmette, a French trader who settled in Chicago in 1790, offered the first known ferry services for hire to cross the river or guide passage through the Chicago portage. Captain John Whistler noted an "Indian Ferry" at the forks of the river, tended by a man living nearby, on his 1809 map of Fort Dearborn (whether this competed with or referred to Ouilmette is unknown). Residents and soldiers from Fort Dearborn also kept boats along the riverbank and at various small piers.

Chicago's first tavern, known as Wolf Point Tavern, built in 1823, was on the west bank near the forks of the Chicago River. In early 1829, Archibald Caldwell, who ran the tavern, strung a rope across the North Branch and left a canoe on either shore, establishing what was referred to as the "grapevine ferry."[4] On the opposite bank, Samuel Miller's tavern competed as the local meeting place. On June 2, 1829, the commissioners of Peoria County issued the first official ferry license in Chicago to Sam-

uel Miller and Archibald Clybourne for the forks of the Chicago River, near Wolf Point. They paid a two-dollar tax and posted a one-hundred-dollar bond to ensure the faithful performance of their ferry service. Local residents had free passage; all others paid half of the rates of the Peoria ferry, as listed:

| | |
|---|---|
| *Foot Passenger* | *6¼ cents* |
| *Man and Horse* | *12½ cents* |
| *Dearborn sulkey chair with spring* | *50 cents* |
| *One-horse Wagon* | *25 cents* |
| *Four-wheeled Carriage drawn by two Oxen or Horses* | *37½ cents* |
| *Cart with two Oxen* | *37½ cents* |
| *Head of neat cattle or mules* | *10 cents* |
| *Hog, Sheep or goat* | *3 cents* |
| *Hundred-weight of goods, wares and merchandise or bushel of grain* | *6¼ cents* |
| *And all other articles in equal and just proportions.*[5] | |

The main landing was on the southeast bank of the river near Lake Street and utilized a scow (a large rectangular, flat-decked, shallow-draft boat) to transfer people and wagon teams. Like the early bridges, these ferries were created out of self-interest and to increase business.

Mark Beaubien assumed the ferry license in April 1831, having posted a bond and purchased a scow from Samuel Miller. A truly colorful character, Beaubien fathered twenty-three children and was a licensed merchant, landlord, storekeeper, and gambler known for racing horses with local Native Americans. The ferry was soon reportedly left unattended, and those wishing to cross had to operate it themselves. By the fall, Beaubien was brought before the Cook County Court and ordered to attend the ferry "from daylight in the morning, until dark, without stopping."[6] Whether Beaubien complied is unknown; however, the first two bridges over the North and South branches the next year put an end to his ferry business.

In 1839 more crossings were needed, and a ferry at State Street was started and praised by the *Chicago Daily American* as a "successful experiment."[7] Using a well-built thirty-by-twelve-foot scow that offered seats for ladies, the ferry connected to a clean, solid platform and walkway at either shore. Voluntarily supported by patrons for reasons of comfort, convenience, and business, it was considered superior to a similar ferry at Clark Street. The City-run Clark Street ferry opened after removal of the Dearborn Street Bridge that same year. The City put up a small ferry room and provided two lamps, but in 1840 a bridge replaced the entire operation.

In May 1842, a City ordinance required all ferry owners to obtain a license. Noah Scranton, brought to court by the City for running the State Street ferry the past two years in violation of the ordinance, was found not guilty. His attorneys successfully argued that a federal ordinance from 1787 established the water-way as a common highway and therefore forever free. Scranton proposed several schemes that would allow him to run the ferry and pay the City's one-hundred-dollar license fee, but no agree-ment could be reached and his ferry was effectively ended. He next constructed a pleasure boat, the *Commodore Blake,* and, along with a Mr. Z. Woodworth, began a lake ferry service called the Chicago and Michigan City Lines.

In 1847 the Common Council ordered a free ferry to cross from Rush Street to the Lake House, a prominent inn on the north side of the river. Ferries were also employed briefly after the flood of 1849 (including Noah Scranton's State Street ferry, which charged a penny per person to cross) until the bridges could be reconstructed. A ferry at 12th Street (now Roosevelt Road) was established sometime later and then replaced by a bridge in 1855.

The first Rush Street Bridge replaced the Rush Street ferry (or Lake House ferry, as it was sometimes known) in 1856. The following excerpt from Gale's *Reminiscences of Early Chicago* well describes the difficulties of using the ferry:

> Of the early ferry, it may be said that it was invariably on the wrong side of the stream; and just as it was making ready to come over for you, probably a little lumber-craft would come in sight, towed by two men in a yawl, and down to the bottom would go the rope that was stretched across the river to be used for propulsion. The yawl would go through with its tow, the skipper responding to the ferryman's cheery greeting as he passed. The craft safely through, Vain Hope would spit on his calloused hand, seize the protruding spokes of the horizontal windlass on shore, and soon the slimy rope, dripping with the ooze dragged up from the river bottom, was in position. Then, grasping it with his wooden pull near the bow, Vain Hope walked slowly to the stern, repeating this movement until the opposite shore was reached; or, standing in one place, would propel from there. Were you in a hurry or in a helping mood, you would lend a hand. And if the passing of some little merchantman detained you, there was ample compensation for the loss of time in the novelty of sqeeing the stranger, and wondering where it hailed from and what its cargo.[8]

This was the last regular ferry crossing for the next half-century. The only exception was the occasional temporary ferry during replacement of a bridge when a temporary bridge was not em-ployed to maintain the crossing.

# ORIGINS OF JURISDICTION OVER THE CHICAGO RIVER AND ITS BRIDGES

For most of the duration of Chicago's history, shifting Native American tribes controlled the land. In 1534, with exploration of the St. Lawrence River by Jacques Cartier and the creation of New France, French influence came to North America; in subsequent decades, this influence pushed west into the Great Lakes area through exploration, expansion of the fur trade, and missionary work. The "Chicagou" portage was "discovered" by Marquette and Jolliet in 1673 and was the starting point for Robert La Salle's expedition to the Gulf of Mexico (where he claimed the Mississippi River, its tributaries, and all the surrounding land for France in 1682).

British influence over Chicago arrived after secession by the French of the lands east of the Mississippi in the 1763 Treaty of Paris, which ended the French and Indian War (known as the Seven Years' War in Europe). Following the Revolutionary War, the British then ceded these lands to the new United States in the 1783 Treaty of Paris. However, the British would continue to meddle in Indian affairs and vie for control of the Great Lakes and its lucrative fur trade for several decades.

The Northwest Ordinance of 1787 established navigational right-of-way on the Chicago River, declaring "the navigable waters leading into the Mississippi and St. Lawrence, and the carrying places between the same, shall be common highways, and forever free, as well to the inhabitants of the said territory, as to the citizens of the United States, and those of any other states that may be admitted into the confederacy, without any tax, import or duty therefore."* The Treaty of Greenville (1795), ending the Northwest Indian War between the United States and a dozen Native tribes (Chippewa, Delaware, Eel River, Kaskaskia, Kickapoo, Miami, Piankashaw, Potawatomi, Ottowa, Shawnee, Weea, and Wyandot), ceded huge tracts of land, including Chicago, and reasserted the right of a free passage through portages and the waterways.

Significant American influence in Chicago, at that time part of the Indiana Territory, arrived in 1803 with the establishment of Fort Dearborn. The fort was built by Captain John Whistler and held a garrison of sixty-eight men. Chicago became part of the new territory of Illinois in 1809. On August 9, 1812, British-supported Potawatomi attacked and killed or captured most of the garrison and settlers. Fort Dearborn was burned to the ground. In 1816, after the end of the War of 1812, the fort was rebuilt, and in 1818 Illinois became the twenty-first state in the Union. On January 15, 1831, the Illinois General Assembly created Cook County, with Chicago as the county seat. Prior to this, Chicago had been moved in and out of nine successive county jurisdictions between 1790 and 1823. With Chicago's founding as a town in 1833, its Board of Trustees, known as the Common Council, had sole responsibility over the next fifty-seven years for the placement, construction, and maintenance of the ferries and bridges on the Chicago River.

*Congress of the Confederation of the United States, *An Ordinance for the Government of the Territory of the United States, North-West of the River Ohio,* passed July 13, 1787, Article IV.

In the 1920s, the Wrigley Company operated a ferry that transported its executives and employees between company headquarters and the train stations. The oldest existing ferry service operating on the river today is the Wendella Boat Company, which began in 1935. The company now operates the bright-yellow Chicago Water Taxis, running regularly from stops at Michigan Avenue, La Salle/Clark, Madison Street, and Chinatown from spring to fall. Recent improvements in water quality, tourism, and sightseeing over the past couple of decades have encouraged a significant expansion in the number of ferries and tour boats operating on the Chicago River.

## Early Bridges (1832–40)

Chicago's earliest bridges were typically bent bridges constructed using four logs to form a square, or *bent,* as it was called. The bents were sunk into the riverbed twenty to thirty feet apart, with the top several feet exposed. Log stringers, usually in pairs, were laid to connect the bents with the shore. This bridge framework then supported a timber roadway laid across the log stringers, usually composed of logs split lengthwise, six inches in diameter by ten feet long. These bridges usually had no railings, and the timber roadway was not secured by any pin or nail. The timbers often had to be reset, and the structures were reported to tremble and shake under every wagon that passed. As one early settler recalled, in the winter of 1833–34, a "splendid team from Detroit" was so startled crossing the bridge they jumped into the river and drowned before anything could be done.[9] Around this time, the boom in population made lumber for home building so scarce that the fifth town ordinance, issued on November 7, 1833, made any person removing wood from the town's bridges subject to a fine of five dollars for each offense.

The town's first drawbridge, built across the Chicago River in 1834, operated like two castle drawbridges that met in the middle. Built by a shipwright and subject to heavy traffic, this "gallows frame" bridge soon became temperamental and had to be removed five years later. Ironically, this first drawbridge, a double-leaf principle, would prove the ultimate bridging solution for the Chicago River with the development of modern bascule designs more than fifty years later.

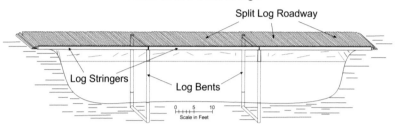

Rendering of wood bent fixed bridge based on descriptions of the first North Branch and South Branch bridges. © Patrick McBriarty

Chicago's first drawbridge, looking east, toward Lake Michigan, in the 1830s.

## Pontoon Swing Bridges (1840–52)

Moveable bridges had become a necessity by the 1840s, and the pontoon swing bridge became the design of choice for the next dozen years. These floating bridges could be built quickly and inexpensively and could accommodate the growing pedes-

## Wooden Pontoon Swing Bridge

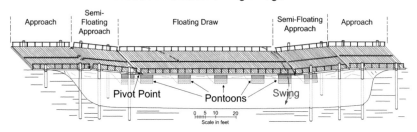

Rendering of a pontoon swing bridge based on specifications for the first Randolph Street Bridge submitted to the Common Council in 1839. © Patrick McBriarty

trian and wagon traffic back and forth across the river. Opening and closing these bridges, however, could take as much as a half hour. Each bridge opened by releasing three corners of the floating draw and pivoting on the fourth corner. It was then pulled to one side using rope or chains connected to the free end and running to either shore. A bridge tender pulled one chain, usually with a windlass, while a second bridge tender slacked. The reverse brought the floating draw back into place to close the bridge. Ships on the river were occasionally known to run over the rope before it sank in an effort to cut it, leaving the bridge agape until the rope could be replaced or repaired; consequently, chain became preferred for its strength and weight, which quickly sank to the bottom of the river, untouched by ships.

By 1849 all of Chicago's bridges at Randolph, Madison, Clark, Wells, and Kinzie streets were pontoon swing bridges. The first pontoons were wood boxes with caulked and pitched seams that created an internal air chamber for buoyancy. Most bridges soon had the added sophistication of incorporating pumps to counter the seepage of water. If not regularly tended, the wood pontoons eventually became waterlogged, and occasionally the weight of a heavy wagon team would push the bridge under-

water. In one attempt to solve this problem, a pontoon made of iron boilerplate was introduced at Wells and Madison streets in 1847. Future designs would soon eliminate this problem, but on March 12, 1849, many docks, most ships, and all of the city's bridges in the Chicago River were destroyed by an early thaw and spring flood.

In 1839 John Van Osdel proposed building a bridge at Wells Street that adopted a railroad-type turntable; it seems, though, that such a bridge was not built there or at any other crossing prior to the flood. Ten years later, all of the bridges built immediately after the flood incorporated a railroad turntable to open and close the floating pier (or draw). The significant mechanical advantage of this relatively new technology allowed the construction of much heavier and stronger pontoon swing bridges. Supported by a circular pier, these float bridges were sufficient to withstand a spring flood in 1857 similar to the one in 1849. These wood bridges typically lasted eight to ten years and were replaced by a more sophisticated design during the next decade. Turntable pontoon swing-bridge designs did not entirely disappear, though, and were often used as temporary bridges to maintain traffic during bridge construction up into the mid-1940s.

## Pontoon Turntable Swing Bridge

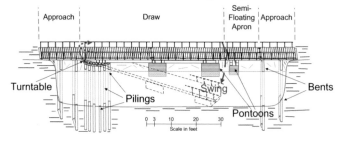

Rendering of a pontoon turntable swing bridge based on specifications for the third Kinzie Street Bridge submitted to the Common Council in 1849. © Patrick McBriarty

The following is transcribed from Rufus Blanchard, *Discovery and Conquests of the Northwest with the History of Chicago*, 565–68.

## THE FLOOD OF 1849

The last thing one might expect in Chicago, situated as it is, on almost a dead-level, is a flood in one of the branches of its river. But this actually took place one fine morning in March, 1849. After a two or three days heavy rain, which had been preceded by hard snow storms during the later part of the winter, the citizens of the town were aroused from their slumbers by reports that the ice in the Desplaines river had broken up; that its channel had become gorged with it; that this had so dammed up its waters as to turn them into Mud Lake; that in turn, they were flowing thence into the natural estuary, which then connected the sources of the South branch of the Chicago river with the Desplaines. These reports proved to be correct. Further, it was also rumored that the pressure of the waters was now breaking up the ice in the South branch and branches; that the branch was becoming gorged in the main channel at various points, and that if something were not done, the shipping, which had been tied up for the winter along the wharves, would be seriously damaged.

Of course each owner, or person in charge, at once sought the safety of his vessel, added additional moorings to those already in use, while all waited with anxiety and trepidation the result of the totally unexpected catastrophe. It was not long in coming. The river soon began to swell, the waters lifting the ice to within two or three feet of the surface of the wharves; between nine and ten A.M. loud reports as of distant artillery were heard towards the southern extremity of the town, indicating that the ice was breaking up. Soon, to these were added the sounds proceeding from crashing timbers, from hawsers tearing away the piles around which they were vainly fastened, or snapping like so much pack-thread, on account of the strain upon them. To these in turn were succeeded the cries of people calling to the parties in charge of the vessels and canal boats to escape ere it would be too late; while nearly all the males, and hundreds of the female population, hurried from their homes to the banks of the river to witness what was by this time considered to be inevitable, namely, a catastrophe such as the city never before sustained. It was not long before every vessel and canal-boat in the south branch, except a few which had been secured in one or two little creeks, which then connected with the main channel, was swept with resistless force toward the lakes. As fast as the channel at one spot became crowded with ice and vessels intermingled, the whole mass would dam up the water, which, rising in the rear of the obstruction, would propel vessels and ice forward with the force of an enormous catapult. Every lightly constructed vessel would at once be crushed as if it were an egg-shell; canal-boats disappeared from sight under the gorge of the ships and ice, and came into view below it in small pieces, strewing the surface of the boiling water.

At length a number of vessels were violently precipitated against Randolph street bridge, then a comparatively frail structure, and which was torn from its place in a few seconds, forcing its way into the main channel of the river. The gorge of natural and artificial materials—of ice and wood and iron—kept on its resistless way to the principal and last remaining bridge in the city, on Clark St. This structure had been constructed on piles, and it was supposed would prevent the vessels already caught up by the ice from being swept out into the lake.

But the momentum already attained by the great mass of ice, which had even lifted some of the vessels bodily out of the water, was too great for any ordinary structure of wood, or even stone or iron to resist, and the moment this accumulated material struck the bridge, it was swept to utter destruction, and with a crash, the noise of which could be heard all over the then city, while the ice below it broke up with reports as if from a whole park of artillery. The scene just below the bridge after the material composing the gorge had swept by the place just occupied by the structure, was something that bordered on the terrific.

The cries and shouts of the people, the crash of timbers, the toppling over of tall masts, which were in many cases broken short off on a level with the decks of the vessels, and the appearance of the crowds fleeing terror-stricken from the scene through Clark and Dearborn streets, were sounds and sights never to be forgotten by those who witnessed them. At State street, where the river bends, the mass of material was again brought to a stand, the ice below resisting the accumulated pressure, and the large number of vessels in the ruck, most of which were of the best class, the poorer ones having previously been utterly destroyed, helping to hold the whole together. In the meantime several canal boats, and in one instance a schooner with rigging all standing, were swept under this instantaneously constructed bridge, coming out on the eastern side thereof in shapeless masses of wreck, in the instance of the schooner, and of matchwood in the instances of the canal boats. Presently the ice below this last gorge began to give way, clear water appearing, while a view out into the lake showed that there was no ice to be seen. It was then that some bold fellows armed with axes, sprang upon the vessels thus jammed together, and in danger of destruction.

Among the foremost and most fearless were: R. C. Bristol, of the forwarding house of Bristol & Porter; Alvin Calhoun, a builder, brother to John Calhoun, founder of the Chicago Democrat newspaper, and father of Mrs. Joseph K. C. Forrest, Cyrus P. Bradley, subsequently Sheriff, and Chief of Police, and Darius Knights, still an employe [*sic*] of the city. These gentlemen, at the risk of their lives succeeded in detaching the vessels at the Eastern end of the gorge, one by one, from the ruck, until finally some ten or twelve large ships, relieved from their dangerous positions, floated out into the lake, their preservers proudly standing on their decks, and returning with salutes, the cheers of the crowd on shore. Once in

THE FLOOD OF 1849.

The Chicago River and Harbor after the flood of 1849, looking east, toward Lake Michigan.

the lake, the vessels were secured, in some cases by dropping the anchors, and in others by being brought up at the piers by the aid of hawsers.

The *Democrat* of the 14th, in its record of the event, says (speaking of the upper jam): "Below all this lies another more solid dam, composed of larger vessels, and consequently stronger material, wedged in so firmly as to defy extraction. . . . Thus is formed on of the most costly bridges ever constructed in the West, and the only one Chicago now boasts of. Crowds of persons were at the wrecks yesterday, and crowded the decks of the various vessels. Many ladies were not afraid to venture over this novel causeway, beneath which the water roared, falling in cascades from one obstruction to another, the whole forming perhaps the most exciting scene ever witnessed here."

## Pivot and Swing Bridges (1852–90)

Pivot bridges, as early swing bridges were known, were designed to replace the pontoon swing bridge. These bridges used new truss designs that began in the United States with the Town lattice-wood truss patented by Ithiel Town in 1820. This was followed in 1840 by a patented wood and iron truss design by William Howe of Spencer, Massachusetts. By the 1850s, pivot bridges were already in use on the Eastern Seaboard. Various Howe truss designs were used by the railroads both nationally and in Chicago on the swing bridges. The flat topography of the relatively narrow Chicago River created an ideal application for the pivot or swing truss bridge designs.

The wood truss gave pivot bridges several advantages: they were raised above the river rather than floating *on* the river, permitting passage of small craft without opening the bridge, allowing for heavier and more durable structures, and reducing bridge maintenance. Pivot bridges typically provided forty- to sixty-foot draws on each side of the center pier for the passage of ships. When closed, the truss structures were supported at the ends by the approaches. In the center of the bridge, a circular pier, usually in the middle of the river channel, held the turntable, consisting of a center pivot surrounded by friction rollers. The superstructure rotated on the pivot like a lazy Susan as a pinion or drive gear connected with the large circular gear. This gave the bridge tenders the mechanical advantage in turning these heavier bridges, which were supported by the center pier and turntable while open. At least a half-dozen pivot bridges were built in Chicago between 1852 and 1858. Most followed the design plans developed by City superintendent of public works Derastus Harper, who constructed the first three of these bridges at Lake, Clark, and Wells streets.

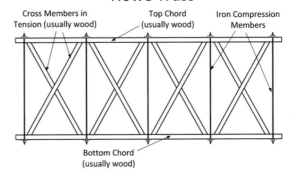

### Howe Truss

Cross Members in Tension (usually wood)   Top Chord (usually wood)   Iron Compression Members

Bottom Chord (usually wood)

A Howe truss. © Patrick McBriarty

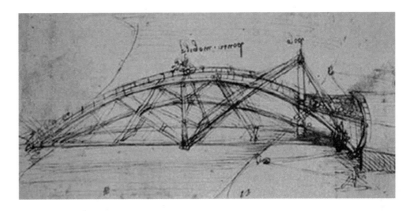

The earliest known drawing of a pivot-bridge design from Leonardo da Vinci's notebooks (ca. 1485–90).

These early swing bridges were far superior to the pontoon floating bridges. They could open and close in a fraction of the time and carried a wider roadway to accommodate more traffic and much heavier loads. The two draws also worked well with maritime navigation rules: approaching vessels were to stay to starboard (to the right side) of the waterway so as to pass port to port (left side to left side). The swing bridge left a channel open in each direction with the open swing bridge between them and so worked well with existing maritime practices.

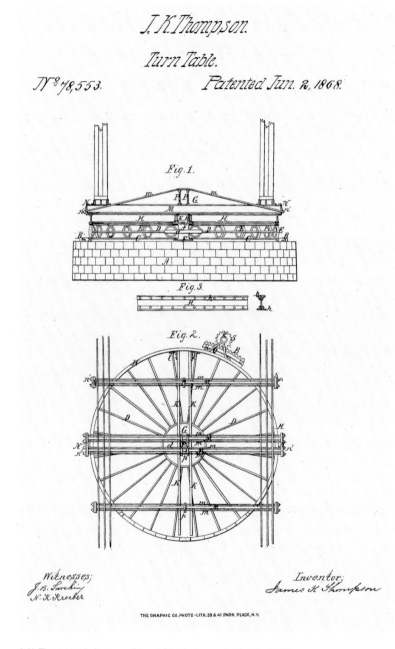

J. K. Thompson's improved turntable design, patented in 1868.

The introduction of rim-bearing turntables provided increased stability for pivot bridges. With this change, the weight of the bridge was carried by the ring of cast-iron rollers instead of being concentrated at the center pivot. Thereafter, both center- and rim-bearing swing bridges were simply referred to as swing bridges. For decades engineers would argue over whether center-bearing or rim-bearing turntables were best, which may have had to do with the different applications for bridges or railroads. Ultimately, the best bridge rim-bearing turntable designs were a hybrid, distributing the weight of the structure evenly over eight points to ensure an even wear of the rollers and on the center pin, allowing the bridge to turn more easily and keep the center pin and turntable in proper alignment.

The first Chicago rim-bearing swing bridge was the Rush Street Bridge, built in 1856. It was followed in 1857 by the third Madison Street Bridge, which utilized two tubular iron-arch bridges patented in the same year by Thomas Moseley of Cincinnati, Ohio. This novel arrangement was never reused, but it further demonstrates that, even early on, Chicago initiated and adopted new bridge technology.

In 1868 Chicago commissioner of public works James K. Thompson received two patents for a modified bridge turntable and an improved Howe truss design.[10] The turntable improvement arranged the bearing circle, cross beams, and bearing beams, allowing the turntable to work more easily and possess greater strength and durability through a more efficient design. Thompson's improved Howe truss design accounted for wood components typically needing replacement well before the iron compression members, arranging the assembly of the members to facilitate repairs. Individual members could now be removed and replaced without having to take the entire span out of service.

Both innovations were incorporated in five city bridges throughout 1867 and 1868. Upon receiving his patents, Thompson petitioned and received a royalty of $250 per turntable and $2 per linear truss foot from the City, totaling $3,028. At that time, it was acceptable for government employees to pursue commercial ventures so long as it did not interfere with their day-to-day work. The City attorney approved the payment, as, prior to construction, Thompson had notified members of the Common Council regarding incorporation of these innovations. Exempting those destroyed by the Great Fire, Thompson's bridges proved their worth and lasted almost twice as long as earlier city swing bridges. This was just one of many refinements and innovations made to the extremely robust swing-bridge design, allowing for construction of longer, wider, heavier spans to convey increased traffic and more formidable loads. Most Chicago bridges built through the early 1870s used some form of a Howe truss superstructure. The swing bridge was the dominant moveable-bridge design both in Chicago and nationally for more than forty years.

The Great Chicago Fire of 1871 destroyed eight bridges at Chicago Avenue and Rush, State, Clark, Wells, Van Buren, Polk, and Adams streets. Seven of these were wood or wood and iron bridges, and the eighth was the iron Rush Street Bridge. A major portion of the city was destroyed, and the fire left more than one hundred thousand people—about a third of the population—homeless. Thereafter, all-iron bridges were preferred. The city's need to reestablish the destroyed crossings was of critical importance, and all eight bridges were replaced in 1872. Five of these eight were all-iron bridges. At least one of these new bridges introduced to Chicago the Pratt truss, invented by Thomas Pratt in 1842. In the late 1870s, the increasing scale of the Bessemer and open-hearth processes made steel more

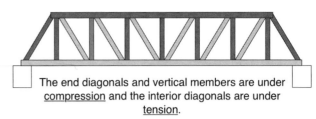

# Pratt Truss

The end diagonals and vertical members are under <u>compression</u> and the interior diagonals are under <u>tension</u>.

# Warren Truss

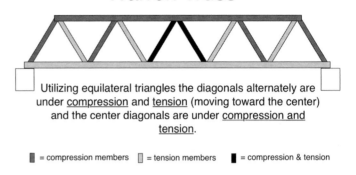

Utilizing equilateral triangles the diagonals alternately are under <u>compression</u> and <u>tension</u> (moving toward the center) and the center diagonals are under <u>compression and tension</u>.

■ = compression members  ▯ = tension members  ■ = compression & tension

Warren and Pratt trusses. © Patrick McBriarty

readily available, of superior quality, and affordable. New all-steel bridges used either a Pratt or a Warren truss superstructure. The Warren truss was patented in 1848 by British engineers James Warren and Willoughby Monzani. After the mid-1880s, most new Chicago bridges were built of steel. Chicago's first steel bridge was the Chicago & North Western Railway Bridge near Kinzie Street and was among the first all-steel railroad bridges in the United States. Both this bridge and the all-steel Glasgow Bridge over the Missouri River for the Chicago & Alton Railroad were completed in 1879.

Still, at this time, most Chicago swing bridges were operated by hand and took ten to fifteen minutes to open and close. The bridge tender used a T-bar or angular iron bar called a "key" that was four to six feet in length and mated with a fitting in the center of the bridge deck. Once in place, the long handle extended to chest height, and the bridge tender pushed it around in a circle to rotate the bridge. The key turned a shaft, and a pinion gear meshed with the large circumference gear under the bridge. By the 1880s, steam-powered engines were incorporated into many of the bridges, and by the mid-1890s, electricity began to replace steam power. Bridges converted to electric power typically used 600-volt direct current supplied by streetcar companies. City agreements granting streetcar companies the right to use the bridges required that power be run to the bridge. Opening or closing a bridge now took as little as forty seconds, about 20 percent faster than steam power. However, it was (and still is) the passage of ships that consumed most of the time during a bridge opening, delaying street traffic and causing congestion.

## Early Regulation of the Bridges

City regulation of the bridges began during the swing-bridge era. In August 1856, a new ordinance required tugboats to "cut down, or lower their smoke stacks, so as to admit of their passage under Wells Street and Randolph Street bridges."[11] This was done so that bridges were not forced to open when the tugs did not have a ship in tow. Most of the tug owners refused to comply, believing the ordinance was arbitrary and that varying bridge clearances made compliance nearly impossible. The ordinance was enforced by the City, though, and soon only a few tugs were able to navigate the river. In response, all tugs refused to tow ships beyond the Clark Street Bridge. For a full day, ships were left to hand-haul their way through the remaining bridges. This led to almost constantly open bridges that caused intolerable delays and congestion and benefited neither party. Over the next few days, the crisis was resolved, with the tug owners agreeing to comply and the number of bridge openings significantly reduced.

A decade later, the Corporation Council studied the rights of the City to regulate the bridges. In April 1867, an ordinance was passed requiring that, after a full ten minutes of delay to street traffic, bridge tenders must then close the bridge and keep it closed for a full ten minutes between 6:00 a.m. and 7:00 p.m. This was challenged by navigational interests, but the City's right to regulate the bridges was upheld by the Illinois Supreme Court in February 1870. Three years later, it was suggested that the city's twenty-nine bridges stay closed during the day and open to ships at night. After careful consideration, the Board of Public Works concluded in its annual report that this "would be a serious blow to the city itself" and that such a rule would be an assault on the "industry and enterprise of Chicago" and in turn injure the working classes who had themselves requested the change.[12]

Navigation season continued pretty much as it had been from mid-April to December, as ships held right-of-way and the bridges had to open on demand. This caused the citizenry and local businesses great consternation and contrasted dramatically with the winter, when the bridges provided uninterrupted crossings. An 1870 survey of traffic revealed Madison Street as the city's busiest bridge, crossed by 44,490 pedestrians and 4,265 vehicles between 6:00 a.m. and 7:00 p.m. each day. Rush Street was the eighth busiest, with 10,385 pedestrians and 2,342 vehicles crossing during the same thirteen-hour period.

In 1874 the bridge ordinance was amended so that between 6:00 a.m. and 7:00 p.m., bridge openings were limited to no more than ten minutes. That same year, a proposal that was tabled and sent to the Bridges and Harbor Committee suggested that bridges carrying streetcar lines stay closed from 6:30 to 8:00 a.m., 12:00 to 2:00 p.m., and 6:00 to 8:00 p.m. In early

1881, the bridge ordinance was again amended (to the outrage of navigational interests) by ordering the bridges not to open between 6:00 and 7:00 a.m. and 5:30 and 6:30 p.m., Monday through Saturday. The ordinance was contested in *Escanaba & Lake Michigan Transportation Company v. City of Chicago,* which reached the U.S. Supreme Court in 1883.[13] The power of the City of Chicago to regulate bridges was again upheld.

The increase in ship traffic saw the busiest bridges opening three to four times per hour during the day. This meant that a bridge might be open twenty to twenty-five minutes per hour during daylight hours. As the Citizens' Association reported in 1884, "Our street-railway and bridge traffic, already enormous, is increasing very rapidly, and the necessity for additional accommodations and facilities becomes daily more apparent. The obstruction to street traffic caused by the bridges, has also become a question of momentous, if not paramount, importance to the business interests of the city."[14] The frustration of the residents was palpable, to the point that commissioner of public works DeWitt Cregier lost his reelection bid in 1886 for protecting marine interests by neglecting to enforce the bridge ordinance.

## Experimentation and New Designs (1890–1902)

Between 1890 and 1902, no other city in the world experimented with as many different moveable and bascule-type bridges as Chicago. Maritime interests demanded better waterways, as the center pier, the ultimate design flaw of the swing bridge, had become a major obstacle on the busy and narrow Chicago River. Shipping was at its peak, and the dramatic increase in the size of steel hulls of oceangoing and Great Lakes ships led to federal oversight over national waterways. With passage of the Rivers and Harbors Act of 1890, Congress authorized the War Department to approve plans for all new bridges over navigable waterways and to seek the alteration of any existing bridge that interfered with navigation. The act also gave the U.S. War Department specific jurisdiction over Chicago waterways, including the Chicago Harbor, the Main Channel of the Chicago River, the North Branch to Belmont Avenue, and the South Branch up to and including the Sanitary & Ship Canal. These efforts were aimed at establishing consistent channel, harbor, and navigational aids across the nation's waterways. The Army Corps of Engineers, which had long ago established a Chicago presence, provided survey expertise and recommendations to the secretary of war. Once reported, the secretary could order removal of obstructions to navigation, which included bridges, and fine local authorities as much as five thousand dollars for every month an obstruction remained. The legislation signaled the eventual end of swing bridges in Chicago.

The City of Chicago's construction of a new swing bridge at Canal Street the same year provided the first test of the new Rivers and Harbors Act. The new Canal Street Bridge was quickly declared an obstruction to navigation by the U.S. Army Corps of Engineers due to the restricted channel and its location near a bend in the river. A legal battle ensued, and the U.S. Supreme Court upheld the new legislation. As a result, the City of Chicago was ordered to remove the bridge just two years after it was built. Subsequently, city bridges were subjected to federal approval and review; however, no viable alternative to the swing bridge yet existed. This created an urgent need for a new moveable-bridge design that would provide an unobstructed channel width of at least one hundred feet, creating a serious problem for the perpetually cash-strapped City. Over the next twenty years, a variety of bridge designs were developed, patented, and built in Chicago and across the nation's waterways in hopes of solving this challenge.

The City of Chicago attempted three different experimental designs over the next decade. The first was a folding-lift bridge, designed and patented by Captain William Harman. The first double-leaf bridge, this nonessential crossing was built cheaply at Weed Street in 1891. Each leaf was double hinged and folded up and back on itself like a jackknife, giving the bridge its nickname. The City, dissatisfied with this first bridge but without any real alternatives, built a second bridge at Canal Street in 1893 amid assurances that the next one would be greatly improved. Ultimately, the design had too many moving parts and was deemed too complicated. Inadequate to the task, it was replaced eight years later.

The City's next experiment was with a vertical-lift design, patented by J. A. L. Waddell and built at South Halsted Street in 1894.[15] Massive towers on each riverbank connected the roadway, the entirety of which raised and lowered like an elevator. When opened, the bridge provided a 155-foot clearance for ships to pass beneath. It was a significant milestone in engineering and steel construction. Though more reliable than Harman's folding-lift bridge, it was unsightly and was soon determined too expensive to build, maintain, and operate.

The third and most promising design was the Scherzer rolling-lift bridge. The City built the first of these in 1895 at Van Buren Street over the South Branch and received forty-five thousand dollars from the Metropolitan West Side Elevated Train Company (or Met). The Met contributed toward the bridge in a deal with the City to build a similar Scherzer four-track elevated-train bridge a half-block north of Van Buren Street. Chicago engineer William Scherzer drew plans for both bridges before his death. The

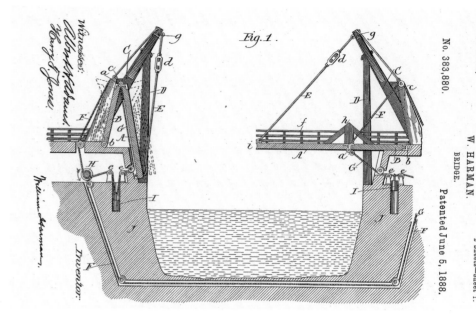

William Harman's folding-lift bridge design, patented in 1888.

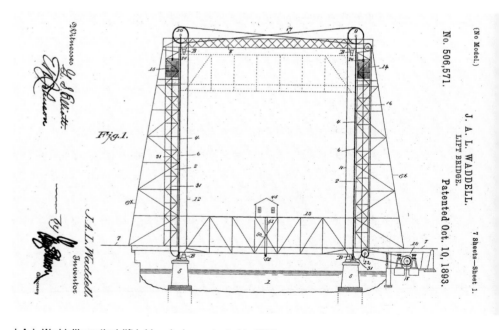

J. A. L. Waddell's vertical-lift bridge design, patented in 1893.

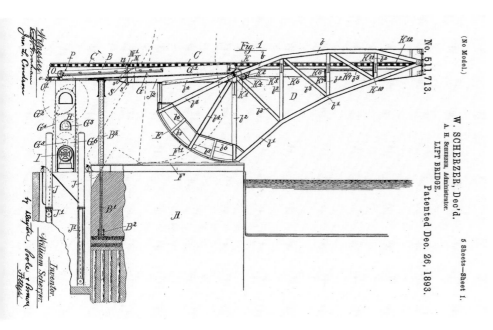

William Scherzer's rolling-lift bridge design, patented in 1893.

A segmental steel girder during construction of a
Scherzer rolling-lift bridge in 1900. Courtesy of MWRD.

Chicago Sanitary District soon favored this design and would construct more than a dozen of these bridges in Chicago within a ten-year period.

These double-leaf bascule bridges operate similarly to a rocking chair. The steelwork of the bridge leaf is built upon two curved, steel segmental girders and opened by rocking up and back along a track foundation. The bridge leaves rose vertically from the water and horizontally back toward the shore simultaneously. The City would build only one other Scherzer rolling-lift bridge, at North Halsted Street across the North Branch, in 1897. City engineers soon discovered a problem with the design. Chicago's geology, composed of eighty to one hundred feet of soil, sand, and clay, provided a poor foundation for such massive moveable structures. It was determined that variable wind pressures during operation of these Scherzer bridges caused the

moveable leaves to shift and wag. The changing pressures resulted in rapid deterioration and subsequent repair of the heavy segmental steel girders. In addition, the movement caused the concrete foundations to shift as much as two inches per year toward the center of the river.

By 1898 City engineer John Ericson, dissatisfied with private-sector bridging solutions, decided the City needed to develop its own design. With little to no budget, the City Bridge Department undertook an exhaustive literature review of European and North American movable-bridge designs. The Sanitary & Ship Canal was also being constructed at this time, and the Sanitary District was given the authority to build bridges. New bridges were deemed necessary to connect existing street and railway lines over the new canal.

In 1899 Ericson lobbied for and was granted involvement in the Sanitary District's bridge-selection process to replace the Harman "jackknife" bridge at Canal Street. Despite strong criticisms by Ericson, a Scherzer rolling-lift design was selected. He then lobbied for the new bridge to have a more substantial foundation, requiring deeper supporting piers that reached bedrock eighty to one hundred feet below. The Scherzer Company refused to change its design, as that would threaten their low-cost advantage. Ericson's continued dissatisfaction created only resentment on all sides. The Sanitary District did not maintain most of the bridges it built and therefore lacked the incentive to back Ericson's objections. As a government agency, the Sanitary District generally had to accept the lowest bid.

In early 1900, Mayor Harrison H. Carter, under federal pressure to replace the obstructive center-pier swing bridges, appealed to the Sanitary District to also build new bridges over the river. The Sanitary District needed the river channel to be widened to two hundred feet and its depth increased to twenty-six feet to provide a river cross-section of forty-eight hundred square feet. This was calculated based on a maximum discharge capacity for the Sanitary & Ship Canal of ten thousand cubic feet per second. The river channel was enlarged to ensure the reversed current on the river was at or below 1.25 miles per hour. This current would protect Lake Michigan and not significantly hinder ship traffic plying the waterway. Until the river was widened, however, the Sanitary District had liability concerns regarding shipping accidents that might be caused by the currents, particularly around the center-pier swing bridges, in the already reversed river. So the Sanitary District agreed to build bridges for the City.

Codified in an ordinance, the City gave the Sanitary District authority to construct eight river bridges without interference.

The Sanitary District would ultimately build ten Scherzer rolling-lift bridges and one Page bascule bridge over the Chicago River. Ownership of each bridge was transferred to the City upon completion, so the ultimate responsibility for maintenance and operation of these bridges fell to the Bridge Department and Ericson. The ten bridges were built at Taylor Street in 1901; Canal, Randolph, State, and Throop streets in 1903; Loomis Street in 1904; Harrison and 18th streets and Cermak Road in 1905; and Dearborn Street in 1907.

The Sanitary District's only experimental design, built in 1902, was the Page bascule bridge at South Ashland Avenue. A former Sanitary District engineer, John W. Page was a partner in the Chicago contracting firm of Page and Shnable, which operated from 1900 to 1908. The firm built this double-leaf bridge to operate using a counterweighted, cantilever approach pier. Four years later, the Chicago & Alton Railroad Company built a second Page bascule design across the South Forth of the South Branch that still stands today. Different from its predecessor, the bridge is a single-leaf bascule and uses a curved rack and counterweighted Warren thru-truss superstructure. Unfortunately for Page, neither design ever caught on.

At about this time, Joseph B. Strauss began developing his own bridge designs.[16] In 1902 he opened the Strauss Bascule Bridge Company with offices at Chicago's famous Monadnock Building. The most iconic of these is his single-leaf, heel-trunnion bascule used by the railroads in the early twentieth century. It is easily recognized by the massive concrete counterweights suspended atop the steel superstructure. Though unsightly, this feature provided the economic advantage of eliminating the substantial excavation and foundation work necessary for the Scherzer rolling-lift or Chicago-type bascule bridge. Its lack of

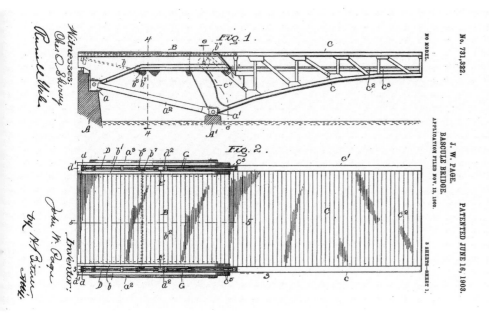

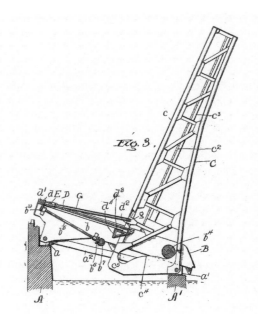

John Page's bascule bridge design, patented in 1903.

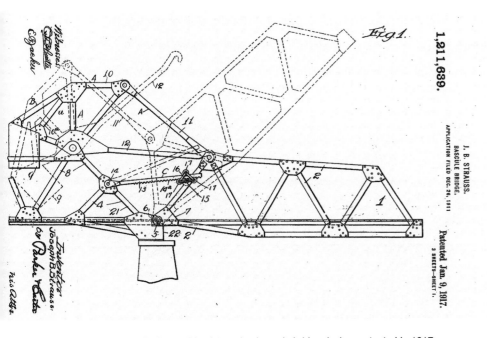

Joseph B. Strauss' heel-trunnion bascule bridge design, patented in 1917.

aesthetics prevented the City of Chicago from ever considering this design, but it was often used by the railroads, and four of these bridges still remain along the North and South branches of the Chicago River. From north to south, these bridges are the Chicago & North Western Railway Bridge, built in 1916 over the North Branch near North Ashland Avenue; the perpetually open and iconic Kinzie Street Railroad Bridge, completed in 1908; and the side-by-side St. Charles Air Line Railroad and Baltimore & Ohio Railroad bridges on the South Branch between 15th and 16th streets. There are also the remains of a Strauss heel-trunnion bridge on the north bank of the Calumet River between 95th Street and the Pennsylvania Railroad vertical-lift bridge.

In 1905 a conference was held between Ericson, Sanitary District commissioner Patterson, bridge engineer Thomas Pihl-feldt, City alderman McCormick, assistant Corporation Counsel attorney William D. Barge, and representatives of the Scherzer

Rolling Lift Bridge Company. It was announced that the foundations were wanting and the bridges too costly to maintain; as a case in point, the State Street Bridge was found inadequate, as the bridge's operation had moved the foundation several inches toward the middle of the river in two years of use. Workers ended up removing two inches from the end of the bridge leaves to prevent overlapping and allow the bridge to close properly. City engineers determined that operation of a Scherzer bascule bridge cost the City an average of $900 more over a six-month period than a similar Chicago-type bascule bridge. According to City engineers, the foundations at the State, Taylor, and Van Buren Street bridges did not meet the required standards. As a result, Patterson declared the Scherzer Company would not be blocked from bidding on City bridge contracts, but would have to ensure that their foundations were made of solid material and that their structures were built more carefully. After 1905 the Sanitary District would complete only two more Scherzer rolling-lift bridges for the City, at Cermak Road and Dearborn Street. The Cermak Road Bridge is the only remaining Sanitary District highway bridge built on the Chicago River.

In 1905, after the election of a new Board of Trustees of the Sanitary District led by Republican Robert R. McCormick, indiscretions, waste, and misuse of funds by the earlier Democratic administrations were uncovered. At the center of the corruption were suspicious amounts paid to a third party, Frank M. Montgomery & Company, for Scherzer design plans. Accusations of graft swirled around the decision to award the Scherzer Rolling Lift Bridge Company the contract to build the eight-track railroad bridges near Western Avenue when the cost amounted to $133,716 more than the next qualifying bid. As a result, the Scherzer Rolling Lift Bridge Company lost its last Chicago customer.

In 1907 Kansas City, Missouri, resident and engineer J. A. L. Waddell's original vertical-lift design, first built at South Halsted Street, was refined in partnership with engineer John L. Harrington. The improvements made by Harrington and Waddell garnered several additional patents, and the vertical-lift bridge developed a following among many of the nation's railroads. In testament to this, the first six miles of the Calumet River are home to six vertical-lift bridges, between Lake Michigan and 130th Street. The Torrence Avenue Bridge, built in 1938, is included in this group and was the second and last vertical-lift bridge ever built by the City of Chicago. More than forty-five vertical-lift bridges were constructed in the United States by 1921.

One vertical-lift bridge remains on the Chicago River, the Amtrak Bridge (originally built by the Pennsylvania Railroad) crossing the South Branch, near Chinatown. Built in 1917, its low clearance requires it to open on demand, and it is controlled remotely from a train-switching tower at 14th and Lumber streets, several blocks away. Free from aesthetic constraints, the railroads preferred patented solutions and often chose the unsightly Strauss heel-trunnion and hulking Waddell-Harrington vertical-lift designs.

Over time, the simple fixed-trunnion designs prevailed over other viable bascule bridges, like the Scherzer or Page design. The Chicago-type and Strauss fixed-trunnion bascules were less expensive to maintain, and the Chicago type avoided the additional cost of patent royalties that incited the ire of taxpayers. This explains why almost all of Chicago's modern drawbridges are Chicago-type bascules.

## The Chicago-Type Bascule

John Ericson and the Bridge Department well understood the need for reliable bridges and the pitfalls of earlier designs by

The Tower Bridge in London over the Thames, which features the double-leaf bascule design.

Bridge Superstructure

Trunnion

Trunnion Bearings

Rotation of Bridge to Open

Side view of trunnion and trunnion bearings of a bascule bridge. © Patrick McBriarty

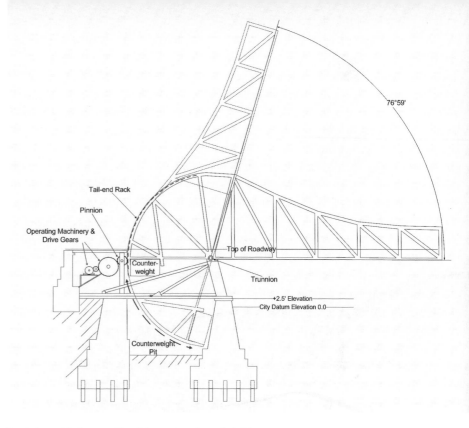

76°59'

Tail-end Rack

Pinnion

Operating Machinery & Drive Gears

Counter-weight

Top of Roadway

Trunnion

+2.5' Elevation
City Datum Elevation 0.0

Counterweight Pit

Rendering of first-generation Chicago-type bascule bridge based on City engineering designs. Drawing by the author.

virtue of having to operate and maintain city bridges. In 1900 the literature review led City engineers to the design that would eventually replace most center-pier swing bridges and early bascule designs. The Tower Bridge in London, England, constructed between 1884 and 1894, was one of the first modern double-leaf, fixed-trunnion bascules. This 260-foot bridge over the River Thames, still the most famous bascule bridge in the world, was well suited to the requirements of the Chicago River. It had few moving parts, was free of patents or royalties, and maintained a fixed load point. This meant that the point of support did not change with the operation of the bridge (unlike, for instance, a Scherzer rolling-lift bridge). The Tower Bridge provided the basis for the Chicago-type bascule developed by John Ericson and the City of Chicago bridge engineers.

In a fixed-trunnion bascule design, the bridge leaf (or leaves) rotates on a fixed point, or trunnion. The trunnion acts similarly to an axle and is held at either end by massive bearings to allow for rotation. With the center of gravity very near the trunnion (or point of rotation), the counterbalanced bridge leaf, weighing as much as several thousand tons, can be operated with a small electric motor. Throughout their history, the Chicago-type bridges have been driven by motors of 60 to 100 horsepower—about the same size as the engine in a Volkswagen Beetle. The initial advantage of the bascule design for Chicago is that it did not need a center support and therefore eliminated the objectionable center pier, providing an open channel of one hundred feet or more.

### First-Generation Chicago-Type Bridges (1902–10)

The first Chicago-type drawbridges evolved from three initial trunnion bascule designs drawn up by City engineers. A Board of Consulting Engineers consisting of Lyman E. Cooley, Bryon B. Carter, and Ralph Modjeski approved of the basic fixed-trunnion concept, though they suggested specific improvements to the substructure, flooring system, and operating equipment. With these recommendations in mind, the City designed and constructed a series of ten bridges that would be known in engineering circles as first-generation Chicago-type bridges. The first-generation Chicago types are identified by their signature high-curve tail ends and three-truss superstructures. They were purely functional in design and did not receive the attention to architecture or aesthetics of subsequent bridges. The first of these was completed at Clybourn Place (later renamed Courtland Street) in 1902.

Indicative of the early development of the steel industry, these bridges were composed of many smaller steel members.

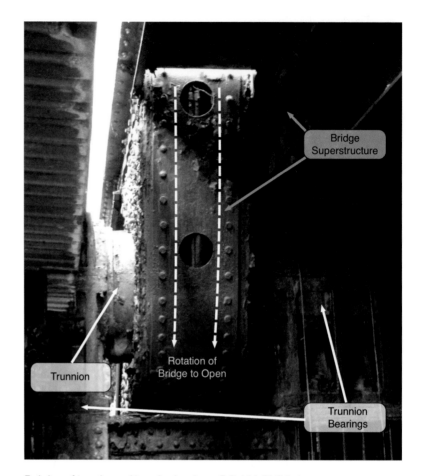

End view of trunnion and trunnion bearings. © Patrick McBriarty

As larger and larger steel shapes and beams became available, they were integrated into subsequent bridges. The first-generation bridges also used cross-bracing to connect the tops of the three trusses. In profile the middle of the structure was lower in height, and the steel angled up toward the ends and then arced down to the roadway at the approaches. Although the trusses were mostly open above the roadbed, the cross-bracing at the approaches led these to be classified as *through-truss* bridges, because traffic traveled through the steel trusswork of the bridge.

The curved end of each truss held a rack that mated with a drive pinion (beneath the roadway), providing the drive mechanism that opened and closed the bridges.

Chicago-type bascule bridges are identified by five basic characteristics:

- counterbalanced truss superstructure
- steel trunnions that provide a fixed point of rotation
- below-deck counterweights affixed under the tail end of each bridge leaf
- electric-powered operating machinery that uses a gear train and rack-and-pinion drive system
- substantial concrete foundations with integral counterweight pits

These bridges are recalibrated by adding to or removing material from the counterweights to keep them in balance. This is done after any changes, repairs, or modifications to the bridge to ensure these graceful structures operate smoothly; even something as innocuous as painting can add several hundred pounds to a bridge and thus require rebalancing. This allows Chicago-type bascules to fully open in less than one minute, which in the case of the first-generation bridges is a maximum angle of 76° 59'.

The first-generation bridges were built with pneumatic buffers to control the motion of the leaf at the end of its travel. Later bridges incorporated gyroscopic controls that automatically slowed the electric motors to control the travel of the bridge as it neared the fully open or fully closed position. The first ten Chicago-type bascules were quickly constructed at a pace of about one per year. Eight were double-leaf bascules, and two were single-leaf bridges (at Kinzie Street and Archer Avenue). The last of these, built at Erie Street in 1910, was radically different from the first nine: it had a higher roadbed with a slight arch

Close-up of the internal rack and pinion of a bascule bridge. © Patrick McBriarty

and used a pony-truss superstructure. A pony truss describes the steel trusswork that extends above the roadbed but has no cross-bracing above the roadway. An internal strut mechanism opened and closed this bridge, allowing for the elimination of the high, curved tail ends of the previous bridges. These design changes at Erie Street foreshadowed what was to come.

## Second-Generation Chicago-Type Bridges (1911–38)

The second-generation Chicago-type bascule bridges were built in two phases: an innovation period from 1911 to 1922 and a refinement period from 1923 to 1938. Although second-generation bridges were designed as early as 1908, a lack of funding delayed

## Through-Truss

Bracing over the bridge deck connecting the trusses
creates a Through-Truss bridge

Traffic flows through
the Bridge Trusses

Traffic flows through
the Bridge Trusses

© Patrick McBriarty

## Pony Truss

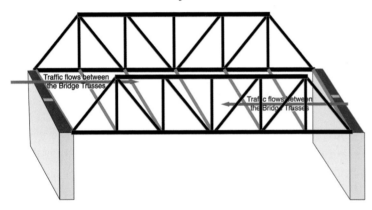

Traffic flows between
the Bridge Trusses

Traffic flows between
the Bridge Trusses

© Patrick McBriarty

## Deck Truss

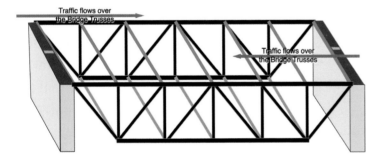

Traffic flows over
the Bridge Trusses

Traffic flows over
the Bridge Trusses

© Patrick McBriarty

construction until the fall of 1911. Alexander von Babo, a City engineer specializing in bridge design, introduced the most important of the early second-generation innovations: the internal rack-and-pinion system. Awarded a patent in 1908, this system became integral in the use of pony-truss superstructures, encouraging bridges with lower profiles and a more graceful appearance.[17]

Another improvement was the increase of the size of the counterweight box to allow for the use of less expensive materials, such as stone or concrete, instead of iron or lead. Other improvements during this period focused either on the support and configuration of the operating machinery or on providing a greater sense of architecture for the bridges. This latter improvement was influenced by the City Beautiful movement, which encouraged public works that beautified the urban landscape for the common good, promoted civic and moral virtue, and improved the quality of life. As this period progressed, city drawbridges became more attractive, displayed greater ornamentation, and integrated the bridge houses architecturally into the overall design of each bridge. Before this, bridge houses were stuck to the side of bridges and looked more like an afterthought than part of the overall design.

From a technical standpoint, the double-deck bascule bridge was one of the greatest innovations made during this period. It solved a serious problem facing the City by allowing for the replacement of two obstructive center-pier swing bridges. Years earlier an upper deck was added to the Lake and Wells Street swing bridges to carry elevated trains into and out of the Loop. (The Loop refers to Chicago's historic commercial center in the heart of downtown and originates from the elevated-train railway loop bounded by Madison, Wabash, State, and Lake streets.) City engineers downplayed the achievement, but the creation of the

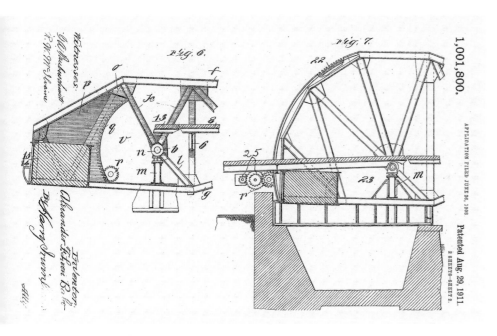

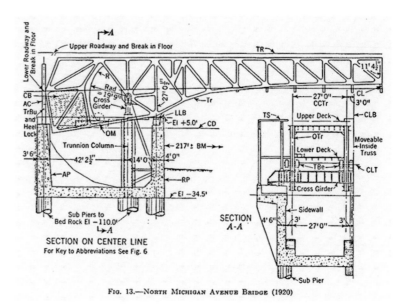

Design drawing of the double-deck Michigan Avenue Bridge, built in 1920.

Comparison of the first-generation rack-and-pinion design (*left*) with von Babo's internal rack-and-pinion design (*right*) patented in 1911.

double-deck bascule was a formidable and technically challenging solution to a very specific Chicago problem. Equally amazing was the fact that elevated-train service was continued for commuters while replacement of these two bridges was completed. Each installation interrupted commuter service on the first bridge for one week and on the second for a single weekend. These amazing feats are detailed in the Lake and Wells Street bridge chapters.

The innovation of a double-deck bascule at Lake Street was also applied to Chicago's iconic Michigan Avenue Bridge. Completed soon after World War I, this new span was the world's most unusual bridge and has become an icon synonymous with the City of Chicago.

Another improvement made during this period was the design of the railing-height truss, first introduced at the Madison

Street Bridge in 1922. It was so called because the top of the steel trusswork ends four feet above the roadbed, at about the height of the sidewalk railings. This provides for a clean appearance and offers the added safety advantage of physically separating pedestrian and street traffic.

The end of the innovation period was signaled by the City's unexpected loss of a patent-infringement suit in 1920. A federal appeals court ruled that the City of Chicago had constructed ten bascule bridges in violation of a Strauss cross-girder patent.[18] The Strauss Bascule Bridge Company filed the $1 million lawsuit in 1913. It was finally settled in 1920 when the City paid $248,500, financed through the sale of bridge bonds. In 1925 Strauss initiated but lost a similar patent suit against the City of Seattle, Washington.

In the wake of this suit, City engineers developed a "standard" Chicago-type design that avoided all existing and potential patent infringements. The Bridge Department created an S-shaped box girder to support future bridge trunnions and avoid the Strauss patent. The standard design also used a new internal rack-and-pinion design, thereby avoiding Alexander von Babo's patent, despite the fact that he had never approached the City for compensation. Creation of this new standard design began the refinement period of the second-generation Chicago-type bridges in 1923. The refinement period would yield a series of design variations that solved specific Chicago River applications; in other words, the new standard design was modified to accommodate site-specific requirements as they were encountered. Designs included specifics dictating the architectural and aesthetic treatments for the bridges influenced by the Chicago Plan Commission and the Municipal Art Committee, both of which advocated "an improved appearance" of Chicago's bridges.[19]

Other significant refinements made during this period included the open-grid steel bridge deck and a shallow-floor truss system. The open-grid steel bridge deck reduced wind resistance, allowing lighter steel to be used on the bridge superstructure while still meeting or exceeding engineering safety requirements. The shallow-floor truss system was developed specifically for the existing South Halsted Street Bridge that was built in 1934. This design solved the problem of providing a safe approach grade for the street running under a nearby railroad overpass, while still connecting with the bridge deck with enough height to meet the federal river-clearance requirements. The resulting superstructure measured only three and a half feet in depth from the top of the roadbed to the bottom of the supporting steel beneath, surpassing earlier bridges that measured five and a half to six feet from the roadbed to the

low steel. The same plans were used four years later on the South Ashland Avenue Bridge, built in 1938, to take advantage of Public Works Administration funds and avoid the delays that came with preparing new plans.

By this time, bridge construction was a very involved process. As explained by City engineer of bridge design Donald N. Becker in 1929:

*In designing a new bridge, the preparation of the final plans upon which bids for construction are sought represent but a part of the entire work required to bring the project to that stage. Surveys must be made, grades and street lines studied, property damages weighed, interference with present utilities minimized and the entire relation of the new bridge to other existing, proposed or even remotely possible structures viewed with regard to usefulness, harmony and economy. This requires consultation with numerous other organizations, both public and private, such as the Federal Government, the Sanitary District of Chicago, the Division of Waterways of the State of Illinois, the Chicago Plan Commission, the Illinois Chapter of the American Architects, the Municipal Art Commission, the Illinois Society of Architects, about a dozen utilities companies, in some instances railroad companies, and various municipal departments. The preliminary work involves the preparation of many studies and auxiliary drawings, calculations and estimates, many of which are of no further use after a decision has been based upon them.*

*After the contracts for construction work have been awarded, shop plans must be checked and a certain amount of supervision exercised and auxiliary plans made as the work in the field progresses.*

*Occasionally foreign drawings are submitted for approval by railroad or other agencies and in such cases the design is checked and the details examined.*[20]

The 1930 Department of Public Works annual report detailed $20 million in future bridge projects. Given financial considerations and the practicality of undertaking all these projects at one time, however, a bond issue of just $11.6 million was requested. The report's top priorities were replacement of the South Halsted Street, State Street, and Torrence Avenue bridges, along with raising funds for the improvement of existing bridges. Improvement projects included the installation of new bridge floors at the Grand and Chicago Avenue, Dearborn, Division (over the canal), 18th, North Halsted (over the canal), Harrison, 22nd, and Van Buren Street bridges. New bridge houses were proposed for the Loomis, Dearborn, Randolph, 95th, Kinzie, Harrison Street, and Western Avenue bridges. Upgrading the mechanical equipment on eighteen bridges—installing new center locks and more substantial motor breaks and heal locks—was planned, along with new barriers and traffic and safety signals for the Franklin and Monroe Street bridges. Installation of electric interlocks on twenty-eight bridges was also proposed. This last item would ensure that all fifty-four of the City-operated drawbridges had systematic roadway signals, heel locks, center locks, and operating machinery that safeguarded bridge operation. Clearly, the responsibility for designing, operating, and maintaining the city's network of bridges was not a trivial undertaking.

## Postwar Chicago-Type Bridges (1948–67)

Construction of postwar Chicago-type bascules ushered in a more austere modern appearance for bridges and bridge houses.

During this era, one-man operation was introduced with the North Halsted Street Bridge in 1955. Self-adjusting roller bearings were also introduced, and open-grate flooring became common. Throughout the 1950s and 1960s, many existing bridges were also converted to one-person operation. Previously, at least two bridge tenders were necessary, one in each bridge house, to safely operate a bridge; later technologies made it possible to remotely control and safely operate a bascule bridge from a single bridge house. Another new invention, closed-circuit television, was also used at several bridges to improve visual checks after conversion to one-man operation. These conversions significantly reduced the annual operation expenses by eliminating many permanent bridge-tending positions.

Although several of these bridge-construction projects were started prior to World War II, all were halted until after the war. This resulted in a seven-year hiatus in bridge construction, the sole exception being the modification of the South Western Avenue Bridge crossing the Sanitary & Ship Canal. In 1942 this bridge was converted from a fixed span to a vertical-lift bridge. Impetus for this project came from the U.S. Navy, which wanted to build ships in the Great Lakes during the war and be able to send them down the Mississippi River to the Gulf of Mexico. All other Chicago bridge projects either were halted by the War Production Board or languished for lack of materials, particularly steel.

After the end of the war, the City completed several projects, including the State Street Subway tunnel and bridge project. This was the first of seven new Chicago-type bridges constructed to replace six early Scherzer rolling-lift bridges and one first-generation Chicago-type bridge at 95th Street. Other postwar bridges came in the form of two new river crossings: the twin drawbridges at Ohio Street and the Congress Parkway, connecting downtown with the new Kennedy and Dan Ryan expressways.

## Modern Chicago Drawbridges (1978–Present)

Only three modern bascule bridges have been built since the postwar era. These are located at Loomis Street, Columbus Drive, and Randolph Street. All incorporate the one-piece box-truss technology first prototyped at Loomis Street in 1978. Improved welding methods and materials allowed the reliable manufacture of large structural steel trusses capable of supporting bridge roadways. The one-piece welded box trusses eliminated the thousands of rivets necessary for the construction of earlier bridges. The huge box trusses were shipped to Chicago and assembled into bridges at the construction sites. This method took drawbridge construction to a whole new level of precision, since with only one point of adjustment, the two bridge leaves had to fit together within fractions of an inch. This design was highly successful and gave modern bridges their clean, streamlined appearance.

## Yes, Virginia, Chicago Still Has Drawbridges!

Today's extensive network of bridges allows residents to hardly give crossing the river a second thought. Several generations removed from the regular passage of commercial ships through downtown, our collective understanding and appreciation for Chicago as a city of drawbridges has dwindled and almost vanished. The opening of the drawbridges for sailing and propeller ships several times an hour, and the resulting congestion of people, horses, wagons, and carriages, now seems otherworldly.

These paradoxical structures of permanence and movement that have carried the hustle and bustle of the city across the river for more than a century are an especially fascinating element of Chicago. The genesis of these precious structures evolved from Chicago, a city that is constantly remaking itself. The drawbridges are true gems on a string of waterways. The dual pressures of land travel and navigational right-of-way forged the city's bridges, as engineers attempted to meet the changing demands of both.

Chicago first built float and swing and then bascule bridges, each type cleverly designed to be all the rage. In time each archetype became passé and was replaced by the next design. As Chicago matured, the bridges became more refined, more costly, and more valuable. Bridge materials and know-how progressed, and, as a center for drawbridge technology, Chicago attracted builders, engineers, and architects, the jewelers of drawbridge design. Only Amsterdam has more moveable bridges than Chicago, ironically also known as the Second City, but no other city on earth developed as rapidly, developed as many, built such a variety, or used drawbridges as intensively, making Chicago the drawbridge capital of the world. This attracted many of the top bridge engineers with names like Scherzer, Strauss, Page, Pihlfeldt, Waddell, and Ericson and advanced the cutting edge of drawbridge technology and design. At its peak in the early twentieth century, Chicago was among the best places an engineer could leave a mark on the changing urban landscape.

After one hundred years, the dominance of moveable bridges began to be undercut by the changing trends in shipping, industry, and urban development. The use of drawbridges on the Chicago River had peaked. Improvements, refinements, and new technologies certainly continued, but at a much slower pace. The Chicago River had long since lost its crown as the world's busiest harbor. The bridge clearances and standardized designs easily accommodate today's low-slung commercial barges, tourist boats, and most private boat traffic without needing to be opened. The real demand for drawbridges is a fraction of what it once was.

It is expected that the City of Chicago will continue to gradually restrict or discourage navigational rights on the Chicago River in order to cut costs. The drawbridges are no longer required to open on demand and are lifted just a few dozen times each spring and fall. These scheduled lifts accommodate the passage of recreational sailboats to the few remaining boatyards on the Chicago River, yet even this traffic has diminished; in 2004, Crowley's, the city's largest boatyard with winter storage for more than eight hundred boats, moved from the Chicago River to a larger facility on the Calumet River, near 95th Street.

The river itself has also changed. Once not much better than an open sewer for residential and industrial waste, the Chicago River has now seen the return of fish and wildlife. Gone are most of the lumberyards, grain silos, and heavy industrial companies that valued riverfront property for shipping. Likewise, the shipbuilding industry that contributed so much to Chicago's development is now gone. Neighborhoods along the Chicago River, particularly over the past three decades, have been gentrified and are now composed mostly of new or rehabilitated residential and office space. Recreational use of the river is on the rise.

Will Chicago remain the drawbridge capital of the world? Many drawbridges are more than fifty years old, and others are approaching or have surpassed one hundred years of age. Their replacement will be expensive, as the cost of a new bascule bridge is several times that of a fixed span. Given the prevailing conditions, a moveable bridge, when compared to a fixed bridge in terms of application and cost, seems impractical and unjustifiable. On the North Branch, the fixed bridge has supplanted the drawbridge, and replacement of moveable bridges with fixed spans will continue as funds become available. The hundred-year-old drawbridges at Division Street are likely next in line for removal and replacement. Is it any wonder that a new bascule bridge has not been built in Chicago since 1984? As fixed bridges progress down the North Branch, it remains to be seen whether they will migrate to the South Branch and the Main Channel of the river. In Chicago only the Calumet River still hosts large commercial ship traffic, and the bridges still operate on demand, seven days a week, twenty-four hours a day.

Understanding this does not necessarily make bridge fans (also known as pontists) feel any better. The intricate, well-built drawbridges of our industrial past continue to age. Just as the Chicago-type bascule at North Avenue was replaced in 2008 by a new, wider, fixed-suspension bridge. The century-old North Halsted Street bascule over the canal was likewise replaced in 2012 by a new arch suspension bridge. Although aesthetically pleasing, these new fixed bridges are much less fascinating. For some it feels like a betrayal, the trading of true gems for costume jewelry. Only federal oversight, the great cost of replacement, and the robust design of the existing bridges have allowed for the preservation of so many classic and mechanically and architecturally significant spans.

In just under two hundred years, Chicago grew from a swampy western outpost to the metropolis of today. The city's bridges, and particularly its drawbridges, have been an integral part of this growth. Chicago still holds the greatest variety of moveable bridges in the world, a fact that should be appreciated as the historic older drawbridges become threatened by replacement. It is hoped this history will stir the imagination and encourage Chicagoans and visitors alike to now more fully appreciate the world's greatest working drawbridge museum.

# THE BRIDGES OF THE

## (FROM EAST TO WEST)

Map of the Main Channel of the Chicago River and bridges. © Chicago CartoGraphics

Franklin-Orleans
Chicago bascule 1920–

Wells St
Chicago bascule 1922–
double-deck

LaSalle St
Chicago bascule 1928–

Clark St
Chicago bascule 1929–

Dearborn St
Chicago bascule 1963–

State St
Chicago bascule 1949–

Wabash Ave
Chicago bascule 1930–

Rush St
swing 1856–1920

Michigan Ave
Chicago bascule 1920–
double-deck

Columbus Dr
Chicago bascule 1982–

Illinois Central RR
swing 1880–1905

Lake Shore Dr
Strauss bascule 1937–
double-deck

The Main Channel extends one and a half miles from the mouth of the Chicago River at Lake Michigan west to the forks, where it splits into the North and South branches. There has never been a fixed bridge across this section of the river, and federal oversight still requires moveable bridges on the Main Channel of the Chicago River.

Originally, the normally slow current of the Main Channel flowed east into Lake Michigan. At Michigan Avenue, it turned south, meandering the last half mile through the lakefront sand dunes before reaching the lake at Monroe Street. The soldiers of Fort Dearborn dug a channel straight through the sand to open the river in 1818, 1822, and 1829, as sand pushed south along the lakeshore by storms and currents repeatedly filled or threatened to close the new channel. In 1833 Congress appropriated twenty-five thousand dollars for the construction of piers to protect the entrance to the river, reaching out into Lake Michigan to create the Chicago Harbor. This permanently opened the

# MAIN CHANNEL

river mouth as a harbor in 1834, under the supervision of what became the U.S. Army Corps of Engineers (the Corps). Time and again, these piers were extended farther east out into the lake to prevent sand bars from forming that could obstruct the river entrance. Over the decades, the sand built along the north side of the pier gradually pushed the lakefront eastward. This new land, north of the river and east of Michigan Avenue, became what is now known as the Streeterville neighborhood.

The Main Channel was originally 200 to 240 feet wide at the river mouth and varied in depth from 3 to 7 feet. The Corps oversaw improvements creating a natural port and safe harbor for Great Lakes shipping at the southwest corner of Lake Michigan. Chicago's busy commercial traffic made the banks of the river a haven for warehouses, grain elevators, slaughterhouses, lumberyards, and rail lines. The Chicago Harbor and entrance were further improved with the completion of a lakefront harbor in 1893; Municipal Pier (or Navy Pier) was added in 1916, and the surrounding outer harbor break walls were added about the same time. The Corps still oversees harbor improvements and manages the locks at the mouth of the river today. The locks were added in the 1930s to regulate the flow of water after reversing the Chicago River in 1900, which then flowed west into the South Branch and, ultimately, down into the Gulf of Mexico.

The Main Channel of the relatively short, narrow Chicago River became one of the most important waterways in the country and the world. Its development went hand in hand with the growth and development of Chicago. Over the past forty years, business along the river has transitioned from railroad, manufacturing, and warehouses to office and residential space. The formerly polluted river is gaining popularity for recreation and tourism, and buildings no longer turn their backs to the river. Not surprisingly, the Main Channel features the most splendid and decorated bridges on the river, celebrating its original role as a harbor and the gateway to Chicago.

# LAKE SHORE DRIVE BRIDGES

LOCATION: 400 East, 402 North; Lake Shore Drive runs north and south along the lakefront. It crosses the Main Channel of the Chicago River just inside the locks separating the river from Lake Michigan.

HISTORICAL HIGHLIGHT: Originally called Leif Erickson Drive and also Field Boulevard, this most easterly Chicago street was officially renamed Lake Shore Drive in 1946. When it was built in 1937, the Lake Shore Drive Bridge was the longest and widest bascule bridge in the world. It was initially referred to as the *Link Bridge* because it became the busiest connection between the northern and southern sides of the city.

President Franklin D. Roosevelt dedicated the Lake Shore Drive Bridge in October 1937. This 331-foot, double-deck, double-leaf bascule bridge was designed by the Strauss Engineering Corporation in collaboration with Hugh Young, chief engineer of the Chicago Plan Commission. It provides a 220-foot draw for ships, and its 108-foot-wide roadway provides four northbound and four southbound lanes of traffic. This project included a single-leaf bascule bridge one block north over the Michigan Canal, also known as the Ogden Slip. At the opening ceremony, Roosevelt gave his famous Quarantine Speech an-nouncing the policy shift away from isolationism toward greater international responsibility and insisting on the quarantine of aggressor nations.

Lake Shore Drive sits atop a landfill that extends into the lake as much as a half mile. Extension of the city's lakefront was begun by the Illinois Central Railroad in the 1860s. The IC built a break wall several hundred yards into the lake to protect its lakefront rail lines. The resulting lagoon was later backfilled, much of it with rubble removed from the city after the Great Fire of 1871. The *Plan of Chicago* in 1909 provided a vision for ex-tensive development of the city lakefront, including the addition of several lagoons and islands.

The plan to develop Lake Shore Drive started in 1926 as part of a huge lakefront improvement project headed by the Chicago Plan Commission. Lake Shore Drive was included in Burnham and Bennett's *Plan of Chicago,* connecting North and South Lake Shore Drive. The intent was to relieve downtown traffic congestion, enhance the city, and connect Grant Park to Lincoln Park. It involved the input, planning, and cooperation of at least a dozen major stakeholders, including the City, the State of Illinois, several federal agencies, Park Commission represen-tatives, the Department of Public Works, the Sanitary District, the Commercial Club, and the area's two major landowners,

| | | | **Lake Shore Drive Bridge** | | | |
|---|---|---|---|---|---|---|
| Opened | Bridge type | Designed by | Constructed by | | Cost | Status |
| Oct. 19, 1937 | Strauss double leaf, double-deck bascule | Strauss Engineering Co. & Hugh Young | Ketler-Elliott Co. (construction), American Bridge Co. (steel fabrication) | | $11.5 million | Currently in use |

The Lake Shore Drive Bridge, looking west in 2009 with Trump Tower near completion. © Kevin Keeley

the Illinois Central Railroad and the Chicago Canal and Dock Company. This $14 million Great Depression–era project was mostly financed through the sale of park bonds to the federal government, and PWA funds made up the final $2.3 million.

During this time, Chicago hosted the World's Fair in 1933, called "A Century of Progress." Preparation for this exposition helped drive the creation of Northerly Island and further cleanup of the lakefront. Afterward, most of the fairgrounds were retained as public parks, yet the Illinois Central would not relinquish right-of-way or portions of land along the city's lakefront dating back to the 1850s (and secured for the railroad by a young attorney named Abraham Lincoln). As a result, two 90-degree turns connected the new bridge with the northern portion of Lake Shore Drive. These turns became known as the Lake Shore Drive S curve.

The final plan for the lakefront created a symmetrical Monroe harbor, anchored by the Shedd Aquarium on the south side and a similar curve of land to the north. Earlier proposals included a high fixed bridge that would provide a 125-foot clearance and a tunnel under the river. It was determined, however, that building two bascule bridges—a double-leaf bridge over the Chicago River and a single-leaf bridge over the Michigan Canal—would be the most cost-effective plan.

The Michigan Canal is a half-mile slip cut out of the lakefront parallel to the river a few hundred yards north. It was once also known as the Ogden Slip due to William B. Ogden's oversight and investment in the Chicago Canal and Dock Company. The company held considerable property on Chicago's North Side and operated out of the Pugh Terminal at 435 East Illinois Street. A single-leaf bascule crossing the slip accommodated commercial ship traffic. This building was later renovated into retail and commercial space in 1990 and renamed North Pier.

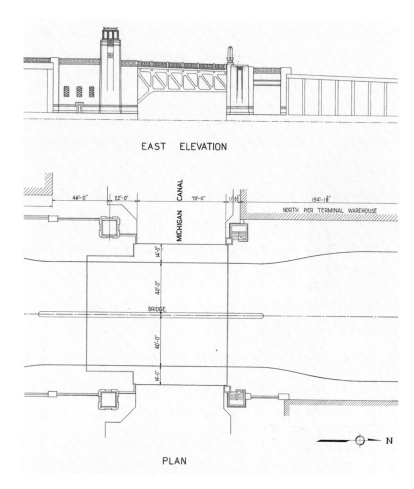

EAST ELEVATION

PLAN

Design drawing of the single-leaf bascule bridge crossing the Michigan Canal from 1937 to 1983. Courtesy of CDOT.

Both bridges were originally designed to carry automobiles on the upper deck and train traffic on the lower deck. Trains were never actually used on either bridge, however, and the lakefront train lines south of the river were later depressed to provide a more open park system.

From 1982 to 1986, Lake Shore Drive was reconfigured to remove the S curve. This major road project included the con-

Looking west at the cluster of bridges and ramps
crossing the Michigan Canal. © Laura Banick

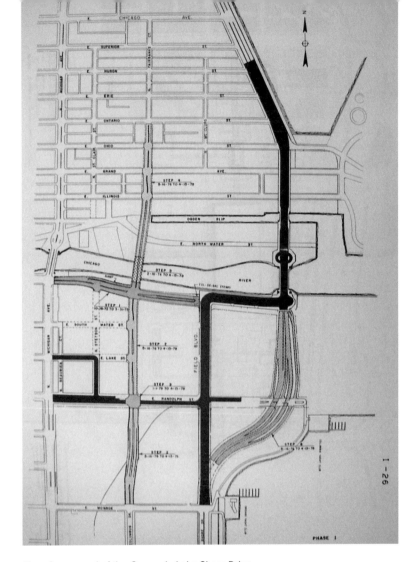

Plans for removal of the *S* curve in Lake Shore Drive
completed between 1982 and 1986.

struction of the Columbus Drive Bridge, which opened in 1982, and tied Randolph Street, Wacker Drive, and Grand and Illinois avenues into Lake Shore Drive. The project also replaced the single-leaf bridge over the Michigan Canal with a cluster of nine fixed bridges on two levels and finally utilized the lower level of the bascule bridge over the river.

Opening the 12,480-ton double-leaf Lake Shore Drive drawbridge has always been a serious undertaking. Given the amount of traffic on Lake Shore Drive, the Chicago Department of Transportation tries to limit bridge openings to no more than fifteen minutes. A bridge-tending crew of at least ten is employed to ensure a safe, smooth lift operation. The bridge sports four bridge towers, though only two house the controls necessary for a lift. A bridge tender and electrician are posted in the southeast and northwest towers to operate each bridge leaf. Additional personnel ensure that pedestrian and vehicle traffic is off the bridge deck, particularly on the lower level, before a lift. There are strict protocols in place, aided by radio communication among CDOT

The fixed Lake Shore Drive Bridge at Diversey Harbor, looking west. © Kevin Keeley

staff. The actual opening and closing of the bridge usually takes only a couple of minutes, and the passage of vessels typically causes most of the traffic delay.

In 2004 an unusual resident was discovered living under the bridge. Richard Dorsay, a homeless man, claimed to have lived under the bridge for three to four years; however, City authorities estimated the occupation to be closer to three to four months. Dorsay slipped into the bridge superstructure from the lower deck and walked down an I-beam to reach a wood shack he had built into the steel beams of the drawbridge. Tapping into the bridge's 110-volt electrical system with extension cords, he was able to power his television, microwave, space heater, and PlayStation, though not all at once. He was evicted and faced charges of criminal trespass on public property, a misdemeanor. Dorsay was later released into the custody of his father, a resident of Burr Ridge. In the meantime, City workers removed the makeshift residence and welded a plate over the entrance to prevent future squatters.

Besides the two Lake Shore Drive bridges mentioned above, there are three additional Lake Shore Drive bridges, over the Diversey Harbor Inlet, the 59th Street Harbor Inlet (just north of Jackson Harbor), and the Jackson Park Lagoon. All of these fixed bridges are made of steel, concrete, and stone. Daniel H. Burnham and C. L. Strobel designed the 59th Street Harbor Inlet Bridge, which was first constructed in 1892 as part of the World's Columbian Exposition. The current 59th Street bridge is a more permanent version of the original design. On either side of the Jackson Park Lagoon Bridge, sculptural reliefs of a ship's prow, Poseidon, and hippopotamus decorate the stonework.

The fixed Lake Shore Drive Bridge near 59th Street crossing the lagoon, looking east. © Patrick McBriarty

The fixed Lake Shore Drive Bridge at Jackson Harbor crossing the lagoon, looking east. © Patrick McBriarty

# COLUMBUS DRIVE BRIDGE

LOCATION: 300 East, 400 North; Columbus Drive runs north and south. It crosses the main branch of the Chicago River three-tenths of a mile west of the river mouth at Lake Michigan. HISTORICAL HIGHLIGHT: Currently, the Columbus Drive Bridge is the world's second-longest and -widest bascule bridge, exceeded only by the 295-foot Ramón de Carranza Bridge over the Bay of Cádiz in Spain.

The 269-foot-long and 111-foot-wide Columbus Drive bascule is the second-newest Chicago drawbridge. It is a double-leaf Chicago type, and its single-piece box-truss construction is based on the prototype built at Loomis Street in 1978.

The Columbus Drive Bridge fully establishes the box-girder truss as a major construction refinement for modern bascule bridges. Eight trusses, four in each leaf, give this almost quarter-block-long bridge a sleek, streamlined appearance. The trusses were fabricated by U.S. Steel in Pittsburgh and then shipped by barge on the Ohio, Mississippi, and Illinois rivers to Chicago for assembly into the bridge. The box-truss design raised bascule bridge construction to a new level of precision, as the design allows for only one point of adjustment on each truss. Earlier designs used rivet construction and had multiple points of adjustment, whereas the Columbus's leaves had to fit almost perfectly on the first try. Amazingly, the massive leaves of six-lane highway were lowered for the first time nose to nose and were off by less than an inch.

Construction of this bridge went smoothly, but the span has not been without its problems. On April 15, 1983, less than four months after it opened, the bridge had to be taken out of service after a City inspector found cracks in three of the operating gears. Somehow, these forged-steel gears, twenty inches thick and four feet in diameter, had cracked. The drawbridge was opened, severing the connection with street traffic while new gears were manufactured. After eight months, the new gears were installed, and the bridge reopened with little ceremony.

A second incident occurred on May 3, 1984, when the Columbus Drive Bridge balked and refused to open, detaining six sailboats. The binding of the center bolts holding the replaced gears caused the malfunction. The boats were forced to wait for more than seven hours while the problem was repaired. Rerouting traffic created additional headaches for City officials, already in the midst of traffic problems because of the project to remove the S curve from Lake Shore Drive. The new Columbus Drive Bridge offered an important alternative to traffic crossing the river, however, and, since these initial problems, the bridge has provided almost thirty years of reliable service to the City of Chicago.

## Columbus Drive Bridge

| Opened | Bridge type | Designed by | Constructed by | Cost | Status |
|---|---|---|---|---|---|
| Oct. 31, 1982 | Chicago type, double-leaf bascule | Bureau of Engineering, Chicago Department of Public Works | Pashen Contractors, Inc. (general contractor), American Bridge Co. (steel fabrication) | $33 million | Currently in use |

Looking west at the Columbus Drive Bridge in 2009. © Laura Banick

# MICHIGAN AVENUE BRIDGE

LOCATION: 100 East, 365 North; Michigan Avenue runs north and south, crossing the main branch of the Chicago River a half-mile west of the mouth of the river at Lake Michigan.

HISTORICAL HIGHLIGHT: The Michigan Avenue Bridge is the most revered and celebrated bridge in Chicago. This iconic Chicago-type bascule bridge was granted landmark status in July 1991, seventy years after its construction, and in October 2010 the bridge was renamed the DuSable Bridge in honor of Jean Baptiste Point de Sable, Chicago's first permanent resident. The proper French spelling and his chosen legal name is Point de Sable, which is pronounced "du Sable" to confound American spelling.

This double-deck bascule bridge replaced the fourth Rush Street Bridge to become a symbol of the city. Associated with the heart of Chicago, it owes its existence to Daniel Burnham and Edward Bennett's *Plan of Chicago.* The Michigan Avenue Bridge and the development of the boulevard to connect the city's northern and southern sides were a centerpiece of the *Plan of Chicago.* The project widened Michigan Avenue from Chicago Avenue to the river south of Randolph Street.

Prior to this decade-long, multimillion-dollar project to improve and widen Michigan Avenue, Rush Street was the city's major thoroughfare. At the turn of the twentieth century, it was estimated that the Rush Street swing bridge carried fully 50 percent of all north- and south-bound downtown traffic. Before 1920 opening this important bridge crossing created a confusion of horse-drawn wagons, automobiles, and pedestrians that defied description. This condition was aggravated by the surroundings: the South Water Street Market stood on the south bank of the river, and the railhead of the Michigan Central and Illinois Central railroads stood to the east; along the north bank of the river was the railroad depot of the Wisconsin Central and Chicago & Galena Union railroads, and east of that were numerous docks and warehouses, adding to the congestion.

During the first decade of the twentieth century, the need for urban and civic planning was so dire that prominent citizens and influential businessmen took matters into their own hands. By 1906 the 325-member Merchants Club of Chicago enlisted architect Daniel Burnham to create a municipal plan for Chicago. The result was Burnham and Bennett's *Plan of Chicago.* The semipublic Chicago Plan Commission was created to promote the moral upbuilding and physical beautification of Chicago. This included reclaiming the lakefront for the public, improving living conditions, increasing park and public playgrounds, and rationalizing the major transportation arteries of the city.

## Michigan Avenue Bridge

| Opened | Bridge type | Designed by | Constructed by | Cost | Status |
|---|---|---|---|---|---|
| May 14, 1920 | Chicago type, double-leaf, double-deck bascule | Thomas Pihlfeldt, Hugh Young, and Edward Bennett | Great Lakes Dredge & Dock Co. (general contractor), American Bridge Co. (steel fabrication) | $14 million | Currently in use |

Looking west at the Michigan Avenue Bridge in 2009. © Kevin Keeley

British architect Edward Bennett was hired to assist Burnham in developing the comprehensive plan. The plan's three sections outlined Chicago's planning history, planning precedents, and physical, social, and cultural environment; presented a future plan for Chicago; and included a means for promotion to gain popular support for implementation. Mayor Busse appointed Charles H. Wacker to permanently chair the Chicago Plan Commission, a nomination approved by the Chicago City Council on November 1, 1909. Over the next several decades, the Chicago Plan Commission would work toward implementing the many civic improvement and infrastructural projects outlined in the *Plan of Chicago.*

The double-deck Michigan Avenue Bridge's importance to the city is reflected in its design and ornamentation. It features four corner pylons that double as bridge houses, and each is constructed of Bedford limestone and varied metalwork in the French baroque style. Strolling past the Michigan Avenue bridge houses, one cannot help but notice their sculptural relief embellishments. The south bridge houses feature the Fort Dearborn Massacre and the rebuilding of the city after the Great Fire. Sculpted by Henry Herring, they were a gift from the B. F. Furguson Fund. The north bridge houses feature explorers Marquette and Jolliet and the town's early settlers Jean Baptiste Point de Sable and John Kinzie. Created by James Earl Fraser, they were donated by William Wrigley Jr.

The design of the Michigan Avenue Bridge superstructure was the work of the Bridge Design Section, under the Bureau of Engineering within the Department of Public Works. This effort was led by Thomas G. Pihlfeldt, engineer of bridges, in consultation with Hugh E. Young, the engineer in charge of bridge

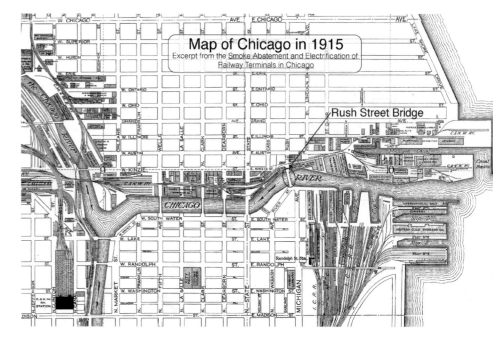

Map of Chicago in 1915 showing the rail terminals surrounding the Main Channel of the river and the Rush Street Bridge.

design. It was designed to resemble the Alexandre III Bridge in France, which was built for the Paris World's Fair in 1900 and is regarded as one of the prettiest bridges in Paris.

This fixed-trunnion bascule is divisible along the center, so it is actually two side-by-side double-leaf bridges. Each half can be raised individually so that half of the bridge (the northbound lanes) may be raised for maintenance or repair, while the other half (the southbound lanes) can carry traffic over the river. This was the first bridge designed to carry automotive traffic on both decks; earlier double-deck bridges carried railway and automobile traffic. The bridge's two upper-deck roadways provide three lanes, each twenty-eight feet in width, and two sidewalks, each

The Alexandre III Bridge over the Seine River, built for the 1900 World's Fair in Paris.

From Huron Street, looking north along Pine Street at the Perry H. Smith house built in 1887 and the water tower in the background.

fourteen feet in width. The lower deck has two lanes going in either direction, each eighteen feet in width. It was the widest bascule bridge in the world when it was built.

On May 14, 1920, the Michigan Avenue Bridge's opening ceremony was nearly the scene of an accident after a lumber steamer, the *Herman H. Hettler,* whistled to pass during the dedication. Navigation having the right-of-way, the bridge tender, not seeing four occupied cars still on the bridge, began to open the southern leaves. Tragedy was averted only after police officers fired their pistols over the din of the crowds and marching bands to get the bridge tender's attention.

The bridge provided a dramatic boon to the city during the twentieth century. More than just relieving congestion and beautifying the city, the bridge spurred real-estate development and economic growth all along Michigan Avenue. Before the bridge, Pine Street (later renamed North Michigan Avenue) was a quiet, tree-lined residential street; the bridge and boulevard brought dramatic changes to create the Magnificent Mile and the Gold Coast neighborhood on the North Side. In the decades that followed, Chicago's "main street" shifted from State Street to Michigan Avenue.

Possibly one of the most fantastic incidents in the history of bridges occurred at the Michigan Avenue Bridge on August 31, 1922. Vincent "the Schemer" Drucci, a known safecracker and member of the Dion O'Bannion gang, noticed two undercover detectives tailing him. Approaching the Michigan Avenue Bridge just as it was being raised, Drucci drove through the warning gates, accelerated up the rising bridge deck, and jumped the gap fifty feet above the river. Police detectives Touhy and Klatzko, not so easily eluded, followed. Racing up the still rising bridge deck, they jumped an even greater gap. Drucci was soon stuck in traffic on the other side of the bridge and fled on foot. Touhy and Klatzko gave chase and made the arrest. To date, this is one of only two bona fide jumps of a Chicago River bridge not staged for a movie. Five years later, Drucci was arrested again; while in custody, en route to the Criminal Court Building, he was shot four times and died during after a scuffle with police officer Daniel Healy.

The Michigan Avenue Bridge provides a good example of the evolution of modern bridge surfaces. Only a few years after its opening, it was determined that the bridge's original four-inch, creosoted wood blocks had become loose and were getting thrown out of place by automotive traffic. In 1924 the old blocks were removed on the southwest section of the upper deck, and a novel one-inch-thick rubber pavement was tested. Interlocking tiles six inches wide by twelve inches long were made from the rubber of old car tires and had a grooved surface for traction. In 1927 the upper deck received one-and-a-quarter-inch rubber-block pavement, atop two inches of tongue-and-groove oak planking. This, along with minor structural repairs, added approximately three hundred tons to each leaf of the bridge.

By 1939 the rubber surface had become badly worn and uneven and required costly maintenance. The old surface was removed, and the six-by-twelve-inch subplanks were reinstalled with bolts using special malleable iron cup washers to dampen the effects of vibrations. Shown above, the subplanks were overlaid by three-inch oak flooring, and topped by one-and-a-half-inch mineral-surfaced asphalt planks over the river. The end of

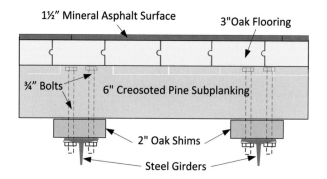

Diagram of the decking system for the Michigan Avenue Bridge in 1939. © Patrick McBriarty

each bridge leaf from the trunnion to the approaches received a concrete-filled one-and-a-quarter-inch-deep iron-grid surface. Experience had shown that traffic entering the bridge would roll asphalt planks into humps and quickly destroy the surface. The concrete-filled grating avoided this problem by acting as a counterweight, allowing for the removal of twenty tons from the counterweight boxes of each leaf. The lower deck was replaced with four-inch paving blocks spaced by one-by-four-inch pine strips and coated with a layer of hot asphalt.

In 1955 both bridge decks were replaced at an estimated cost of $495,000. The upper deck received a concrete-filled steel grid and the lower deck an open steel grid. The unfilled steel grid was used to stay within the weight limits of the bridge structure. In 1973 the Brighton Building and Maintenance Company and the Krug Excavating Company completed a major rehabilitation of the bridge at a cost of $4.5 million. The two-year project required contractors to keep the bridge open to traffic from 6:00 a.m. to 10:00 p.m. on weekdays during replacement of structural members of the bridge viaduct, new steel-grate roadways, widening of the upper roadway, and installation of a new lighting system.

On Sunday, September 20, 1992, the Michigan Avenue Bridge, which was again under repair to install a new roadway, became a gargantuan catapult, hurling equipment and debris hundreds of feet across Wacker Drive. The bridge suddenly sprung upward, sending a forty-ton-capacity crane parked on the end of the southeast bridge leaf tumbling into the counterweight pit. The crane then crashed into lower Wacker Drive and was crushed between the bridge deck and the roadway. The rotating bridge leaf was ripped from its trunnion bearings and plunged to the bottom of the counterweight pit. Vibrations from the closing of the north bridge leaf after the passing of several sailboats may have triggered the incident; however, it was also determined that the heel locks on the south leaf may not have been fully engaged. Several vehicles were damaged, and, most frighteningly, the crane's 285-pound iron ball bounced off Wacker Drive and landed in the backseat of Jesus Lopez's Ford Escort. Lopez, sitting in the front seat at the time, was shaken up, but emerged from the car unscratched. Through the extraordinary efforts of the sixty-five-member construction crew, traffic was resumed across half of the bridge within two weeks, and the full bridge repair was completed a few weeks before Christmas.

Just as the Wabash Street Bridge houses were briefly used for advertising in November 2011, Chicago's bridges have occasionally been used for alternative purposes. During World War II, the southeast bridge house at Michigan Avenue was converted into a recruiting office for the U.S. Maritime Service and processed as many as two hundred volunteer merchant marines per week. More recently, in 2006, the southwest bridge house became home to the McCormick Bridgehouse and Chicago River Museum. A project of Friends of the Chicago River, the Bridgehouse Museum reveals the river's role in the development of the region, highlighting its ongoing transformation from industrial waterway to a natural and recreational asset, and celebrates Chicago's moveable bridges. Visitors can view the bridge's operating gears and counterweights on the lowest level of the museum and, during the spring and fall bridge lifts, watch the machinery in action.

# RUSH STREET BRIDGES

(No Current Bridge)

LOCATION: 80 East, 365 North; Rush Street runs north and south and used to cross the Main Channel a half mile west of the river mouth. There is no bridge at Rush Street today, but the street had four different bridges between 1856 and 1920.

HISTORICAL HIGHLIGHT: The Rush Street Bridge was the most easterly river crossing before Lake Shore Drive, Grant Park, or Columbus Drive existed. The four Rush Street Bridges near the mouth of the Chicago River* were thus collectively nicknamed the *first bridge*. The first three of the Rush Street Bridges were destroyed in succession by cattle, fire, and collision.

The first Rush Street Bridge, built in 1856, replaced a ferry service that was started in 1847. This swing bridge was the first all-iron bridge in the West. A Pratt-truss span, it comprised three trusses supporting two roadways, with sidewalks to each side. One bridge tender and an assistant could operate it from the top deck. Each operator turned T-bars connected to long vertical shafts that ran through the bridge and connected

---

*The exception was the swing bridge at the river mouth constructed by the Illinois Central Railroad in 1880. In December 1903, it was declared an obstruction to navigation and ordered removed by the secretary of war. Removal was completed in June 1904.

with two pinion gears on either side of the turntable. The iron turntable rested between the stone center pier and the bridge superstructure, allowing it to rotate. The octagonal stone and piling center pier was thirty-two feet in diameter and sat in the middle of the river. It comprised wood pilings driven into the riverbed, cut three feet below the low-water mark, and capped with twelve-inch timbers and three-inch oak planking supporting the fifteen-foot-high stone work. Matching pile and stone abutments connected and supported the ends of the bridge in the closed position.

The City of Chicago and the Illinois Central and Galena & Chicago Union railroads each contributed eighteen thousand dollars to the construction of this bridge. It allowed passengers and freight to connect between the IC terminal at the lakefront south of the river with the G&CU freight terminal and Wells Street Station on the north side of the river. This launched Chicago's position as a key transfer point between the eastern and western railroads.

This important river crossing at Rush Street was destroyed on election day in 1863, plunging at least twelve people and about 60 head of cattle into the water. Although it was common practice to open swing bridges while there was still some traffic on the bridge, operating these massive structures demanded sound judgment. With the regular bridge tender away peddling

| First Rush Street Bridge | | | | | |
| --- | --- | --- | --- | --- | --- |
| Opened | Bridge type | Designed by | Constructed by | Cost | Status |
| 1856 | Swing, iron, hand operated | Harper & Tweedale | Harper & Tweedale | $54,000 | Destroyed on Nov. 3, 1863 |

Whitefield's painting of the first Rush Street Bridge from 1860, looking northeast.

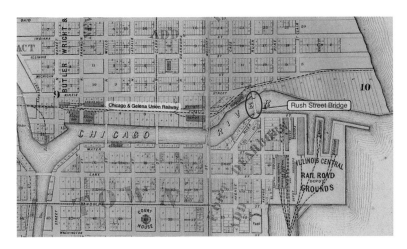

Map of the Main Channel of the Chicago River and Rush Street Bridge in 1863.

tickets for the Copperhead Party, his substitute, characterized by the newspapers as "a very ignorant Irishman," was instructed to swing the bridge whenever a tug whistled.[1]

At a quarter to five that afternoon, the tug *Prindielle,* with two small vessels in tow, whistled for the bridge to open. At the time, two drovers were moving 136 head of cattle across the bridge; a horse and buggy carrying James H. Dole, a wagon team (belonging to Cyrus H. McCormick of International Harvester fame) and its driver, a woman with two children, the bridge tender, his assistant, and four other men were also on the bridge.

Sighting the cattle, the tugboat captain immediately reversed screw and signaled to the vessels in tow to do likewise. Fearing a collision, however, and against the earnest protests of the other men on the bridge and several bystanders, the bridge tender swung the bridge from the abutment. The north end of the bridge, once it left the support of the abutment, began to sink very slowly from the weight of the cattle as it continued to rotate until it had dropped eight or nine feet. The next instant, the south end of

the bridge was elevated twenty feet in the air, and several loud reports were heard from the snapping of iron, the cracking and crash of timbers, and shrieks of horror from bystanders. The Rush Street Bridge had broken in two. It fell across the center pier, leaving hardly a gap through which any ships could pass. Horribly mutilated cattle trapped amid the debris cried out in distress, their bleating mingling with the shrieks for help from the people struggling in the water. Uninjured cattle jostled about in the river, impeding the rescue boats and endangering the lives of those in the water. The approach of dusk further complicated the rescue, creating a dreadfully chaotic scene.

Dole and his horse and buggy were thrown into the river. Both he and one of the cattle drivers, a man named E. D. Gorham, grabbed onto the horns of nearby steers. They alternately floated and sank for about twenty minutes until they were pulled from the river with ropes. The horse and buggy, valued at more than four hundred dollars, were never recovered. Worst of all, one of the children, an unidentified young girl, was never found and

Drawing of the collapsed first Rush Street Bridge in 1863.

The second Rush Street Bridge, looking north along Rush Street, from the third scene of the *Chicago Fire Cyclorama*.

was presumed to have drowned among the cattle and debris. All others were either rescued or picked up by boats or, like one elderly gentleman, aided by a piece of debris, floated to shore. Most of the cows in the river swam to a dock where bystanders pulled them from the water, but two were seen to have drowned, and five more were killed after reaching the dock, their injuries deemed too severe from which to recover. Seventeen head of cattle remained trapped in the wreckage, bleating their distress, which was heard far into the night. With neither of the drovers to be found, the remaining cattle roamed freely through the North Side with no one to tend them.

Years later Gorham, who in addition to being one of the drovers was the owner of the 136 head of cattle, claimed that all were lost. Valuing the animals at fifty dollars apiece, he specu-

lated that approximately half drowned or were crushed to death in the timbers and that the rest were scattered through the city streets. Gorham was quoted in 1883 by the *Chicago Tribune* as saying, "I employed a lawyer to sue the city for damages, but after I had sent $500 of good money after bad the fire came and wiped all out. I wish you'd say to Mayor Harrison that I've always been a good Democrat, and as I am getting a trifle gray I think it would be only fair if the city would, even at this late date, reimburse me for my loss. I'll agree to compromise the matter by accepting half the value of the cattle, and will say nothing about the brutal ducking I got twenty years ago."[2]

Due to the bridge's importance, and to the probable chagrin of Mayor Sherman and the Copperhead Party, the Board of Public Works issued a contract for a new Rush Street Bridge to be built less than two weeks after the accident. Within four months, the second Rush Street Bridge opened to traffic; however, unlike its predecessor, it was of wood rather than of iron construction.

## Second Rush Street Bridge

| Opened | Bridge type | Designed by | Constructed by | Cost | Status |
|--------|-------------|-------------|----------------|------|--------|
| Feb. 3, 1864 | Pivot, wood, hand operated | Fox & Howard | Fox & Howard | $8,740 | Destroyed by fire on Oct. 8–10, 1871 |

The second Rush Street Bridge was a pivot bridge, 211 feet long between the abutments and 32 feet wide, with two 15½-foot carriageways. A two-truss bridge modeled after the Wells Street Bridge (built in 1862), each truss was 24 feet in height and had twenty-six double braces. The bridge deck was composed of Nicholson pavement over 2-inch pine planking.

In September 1864, during a severe gale, the bridge became unmanageable for several moments and was propelled by the wind until it spun around on its axis with great velocity. Several ladies who were on the bridge at the time and even the bridge tender were considerably alarmed by the violent motion. The passengers and bridge survived this unexpected incident, but seven years later, the wood swing bridge would be burned to the waterline during the Great Chicago Fire.

A third Rush Street Bridge was completed in the spring of 1872 as the City of Chicago wasted little time before rebuilding after the Great Fire. This iron Pratt-truss bridge was the same length as its predecessor and a foot wider.

Like its two predecessors, this bridge also met an unexpected end. On November 22, 1883, just before six in the morning, the schooner *Grainger,* laden with lumber and pushed east by the tug *Charles W. Parker,* headed for the south draw of the bridge. According to maritime rules, it should have taken the north draw and by tradition left the bridge to port—or to the left, when facing the ship's bow. As the *Grainger* entered the draw from the east, a steam barge, the *Business,* was entering from the west. As soon as the captain of the *Parker* saw the close quarters, he attempted to get alongside the *Grainger;* though he succeeded in slowing it, the bow was turned toward the bridge and struck port side by the *Business.* The *Grainger*'s jib boom was driven into the bridge, tearing away the boardwalk, one of the iron uprights, and the iron stringer under the bridge.

## Third Rush Street Bridge

| Opened | Bridge type | Designed by | Constructed by | Cost | Status |
|--------|-------------|-------------|----------------|------|--------|
| 1872 | Swing, iron, hand operated | Detroit Bridge & Iron Co. | Detroit Bridge & Iron Co. (superstructure), E. Sweet Jr. & Co. (substructure) | $54,000 | Destroyed by collision on Nov. 22, 1883 |

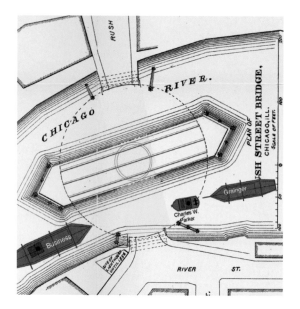

Diagram of the *Grainger* accident as the ships both enter the south draw for the bridge. (*Note:* Ships are not to scale.) © Patrick McBriarty

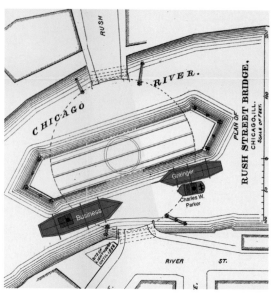

The *Parker* comes alongside the *Grainger* in an attempt to stop a collision. © Patrick McBriarty

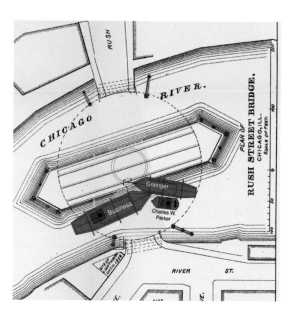

The *Business* collides with the *Grainger,* driving her bow into the Rush Street Bridge. © Patrick McBriarty

This so weakened the structure at its midpoint that it collapsed. Upon inspection, it was deemed better to build an entirely new bridge than to attempt to repair this twelve-year-old structure. The bridge was dismantled and removed, and plans were made for a double roadway bridge that offered two lanes of traffic in each direction.

The fourth Rush Street Bridge was the last all-iron swing bridge the City ever built. At the time it was built, this state-of-the-art, 657-ton Pratt-truss bridge was 240 feet long, 59 feet wide, and the second-largest swing bridge in the world (the largest swing bridge was in Marseilles, France). The old foundation was removed, and an entirely new substructure and new center pier were installed to support the bridge. The center pier was built on white oak piles driven 20 feet into the riverbed and capped 15 feet below the water level by a watertight 2-foot-thick oak caisson. This supported a concrete pier composed of Portland cement, sand, and broken stone, which in turn supported the bridge turntable. The bridge's seventy-four cone-shaped, cast-iron rollers, 18 inches in diameter and 8 inches across their face, were held in place by spider rods radiating from a hub at the center of the 48-foot turntable. The rollers were sandwiched between beveled cast-iron tracks that ensured a smooth, level rotation of the structure. This fourth Rush Street Bridge was also steam powered, and the entire structure was subjected to extensive testing.

Before opening the bridge to the public, contractors ran several tests under the supervision of commissioner of public works Cregier and the bridge's designer, City engineer Artingstall. The

## Fourth Rush Street Bridge

| Opened | Bridge type | Designed by | Constructed by | Cost | Status |
|--------|-------------|-------------|----------------|------|--------|
| July 22, 1884 | Swing bridge, iron, steam powered | Rust & Coolidge | Rust & Coolidge (superstructure), FitzSimons and Connell Co. (substructure) | $125,000 | Closed to traffic on Dec. 22, 1920, and later removed |

bridge operation was timed; it took just 59 seconds to draw the latch and swing the structure in line with the center protection and 62 seconds to swing back to the abutments. Opening the draw for a passing vessel and swinging it back to the abutment for street traffic took 138 seconds. In testing deflection, two City water wagons, each weighing 12 tons, two wagons loaded with

soap, each weighing 12½ tons, and a series of wagon teams with an aggregate weight of 6 to 7½ tons each were driven onto the north end of the bridge. In all, 200 tons of pressure was placed on the north arm of the bridge. This caused only half an inch of deflection on the north end and three-eighths of an inch on the south end. The teams were then moved to the south arm, and

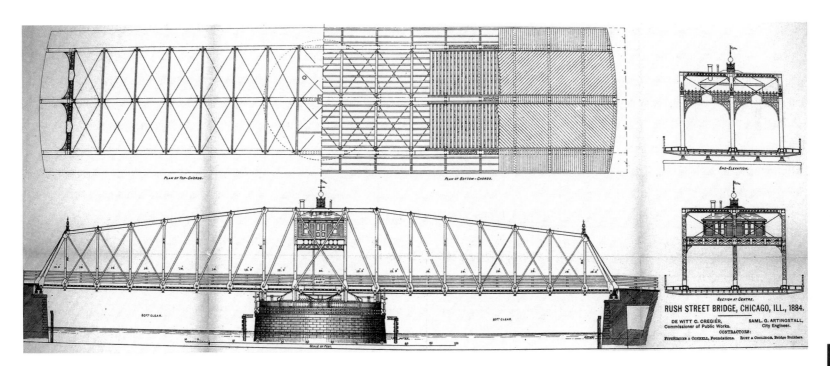

Design drawings of the fourth Rush Street Bridge, built in 1884.

their places were filled by additional teams to cover the bridge from end to end. In total, thirty-four teams and an estimated two thousand people provided a load of at least 370 tons. The maximum deflection of the trusses under this weight was just three-eighths of an inch. Finally, four fire wagons, each pulling a hose carriage, were sent simultaneously at a full gallop across the bridge. Amazingly, almost no jarring was perceptible, and the experts characterized the vibration as moderate. The bridge was swung again, showing no permanent deflection and turning as easily as before.

Even the best bridges require regular maintenance, and this Rush Street Bridge was no exception. In 1895 the steam power plant, a constant source of repair and expense, was removed and replaced by a 20-horsepower electric motor. The worn iron rollers were replaced by a new set of thirty-seven cast-steel rollers. Four years later, the bridge was replanked, one half with creosoted yellow pine blocks and the other, because of a delay in getting creosoted material, with northern pine blocks. In 1902 the northern pine blocks were replaced, and the sidewalks were repaired. In 1904 two new G.E. 800 motors were installed and thereafter were powered by the Chicago Union Traction Company. Previously, the Chicago Edison Company had supplied the electricity. Under this new arrangement, the City reduced the electricity charges for the bridge from one hundred to fifty dollars per month. New maple sidewalks and several new cast-steel sections of rack on the turntable were installed, and new timber bents were constructed to support the south approach. In 1907 danger signals, controlled from the bridge house, were installed

and wired so the bridge could not be opened unless the vibrating gongs and electric lights at the approaches were activated. The following year, the old protection pier was removed, and 468 pilings were driven and received new caps, wales, and braces to complete a new protection pier. This pier was reinforced in 1911 by adding two clumps of 19 pilings at the southeast and southwest corners and a 13-piling clump at the northeast corner. The bridge deck was also repeatedly patched or repaved, in 1911, 1913, and 1916.

The west half of the center pier protection was further improved in 1912. Piling clumps were driven and then an apron of twelve-by-twelve-inch beams was bolted to pilings spaced to form a protection apron around the piling clumps, which was then thoroughly greased to deflect and prevent boats from striking them. This arrangement proved far superior to earlier pier-protection arrangements.

The Michigan Avenue Bridge replaced the fourth Rush Street Bridge after thirty-six years of service. Construction of the Michigan Avenue Bridge, which opened in May, led to shifting of the south abutment of the Rush Street Bridge, forcing its closure to traffic ahead of schedule. On December 22, 1920, the Rush Street Bridge was swung open for the final time; the superstructure was later dismantled and the center and protection pier removed from the river.

The faded relative importance of Rush Street since then illustrates the advantage of having a bridge to a thoroughfare. Had Rush Street received a fifth bridge, the Magnificent Mile might now reside on Rush Street instead of Michigan Avenue.

# WABASH AVENUE BRIDGE

LOCATION: 50 East, 326 North; Wabash Avenue runs north and south, crossing the Main Channel seven-tenths of a mile west of the river mouth at Lake Michigan.

HISTORICAL HIGHLIGHT: The award-winning Wabash Avenue Bridge was built in 1930 and was the longest City-designed bascule bridge in existence for four decades.

The new Wabash Avenue Bridge connected with Cass Street, which was soon rechristened Wabash Avenue north of the river. The bridge was purposely skewed to compensate for the misalignment of the two streets. This, combined with strong navigational demands for a wider channel, resulted in the City's choosing the longest bascule design to date. The bridge provides a clear channel for ships of up to 232 feet and measures 269 feet between each trunnion. The City would not design a longer bridge until the 1970s.

The new bridge relieved the heavy traffic congestion at the Michigan Avenue and State Street bridges. It was the last Beaux Arts–inspired bridge, directly influenced by architect Edward Bennett of the Chicago Plan Commission. Beaux Arts was an architectural style popular in the United States from 1880 to 1920 that was associated with the École des Beaux-Arts (School of Fine Arts) in Paris, from which Bennett graduated in 1902. The style incorporates Greek and Roman architectural features to create massive, elaborate, and often ostentatious effects, and symmetry is used to create vast interior spaces.

City bridge engineer Thomas G. Pihlfeldt designed the Wabash Avenue Bridge, which closely resembles the Roosevelt Road and La Salle, Clark, Madison, and Adams Street bridges built in the 1920s. Pihlfeldt oversaw the designing of all of these bridges except the Clark Street Bridge, which was designed by Paul Shioler. The Wabash Avenue Bridge received the American Institute of Steel Construction's "Most Beautiful Steel Bridge" award for 1930. The jury of engineers and architects cited this span as "a most pleasing solution to a most difficult bridge design problem."[1] Architecturally, this bridge marks the end of an era, as subsequent bridges were less ostentatious in appearance and design and more in line with the modernist movement. Bennett was forced out of office five months before the Wabash Avenue Bridge was completed; the complexion of the Chicago Plan Commission had changed, and his position as consulting architect was abolished.

The Wabash Bridge project included construction of a viaduct over Chicago & North Western rail lines running along the north riverbank. The viaduct was designed to allow construction

| Wabash Avenue Bridge | | | | | |
|---|---|---|---|---|---|
| Opened | Bridge type | Designed by | Constructed by | Cost | Status |
| Dec. 20, 1930 | Chicago type, double-leaf bascule | Thomas Pihlfeldt, Donald Becker, and Edward Bennett | Ketler-Elliott Co. | $2,719,058 | Currently in use |

Looking southwest at the Wabash Avenue Bridge closing during a fall boat run in 2009. © Kevin Keeley

of a bilevel roadway along the north bank of the river to mirror the one recently completed along the south bank. This part of the project, however, was never undertaken.

In 1959 the Wabash Bridge was converted to a one-person operation, and in 1967 it received a full rehabilitation and redecking. On June 5, 1986, the bridge was dedicated to newspaper columnist Irv Kupcinet in recognition of his fifty years with the *Chicago Sun-Times.* The Wabash Avenue Bridge was selected because of its location next to the old *Sun-Times* building on the river's north bank between Wabash Avenue and the Wrigley Building. In 2004 the newspaper relocated, and the old building was torn down. It is now the site of Chicago's Trump Tower.

# STATE STREET BRIDGES

LOCATION: 0 East, 309 North; State Street runs north and south and crosses the Main Channel eight-tenths of a mile west of the river mouth at Lake Michigan.

HISTORICAL HIGHLIGHT: The current State Street Bridge's honorary name is the Bataan-Corregidor Memorial Bridge, dedicated to the Chicagoan World War II soldiers who fought in the siege of the Bataan Peninsula and neighboring Corregidor Island. This attempt to defend the Philippines at the beginning of the war ended in defeat, with many of the approximately thirteen thousand American and Filipino defenders losing their lives in combat. Those who surrendered did not fare much better; about 40 percent of American prisoners of war died in Japanese captivity, many during the Bataan Death March.

After almost two decades of planning and delay, the fifth State Street Bridge was finally completed on May 28, 1949. The double-leaf-trunnion bascule bridge's financing was secured in the fall of 1930 with the passage of a City bond issuance of two and a half million dollars to widen State Street from 66 to 115 feet and three and a half million dollars for construction of the new bridge and approaches. The bridge was given a three-truss rail-height superstructure and, because of the subway tunnel under the river at State Street, heavy steel trusses that straddle the subway and are incorporated into the front walls of the counterweight pits. This structure supports the load of the center truss and then rests on subpiers that reach down to the bedrock on either side of the subway tube. The project, coinciding with various streetcar and rail plans made by the City, was presented to federal authorities as part of a unified system of surface, elevated, and subway lines in the Loop. With the onset of the Depression, it became clear that a special property assessment would be untenable for financing this costly project. The City, working with federal authorities, eventually secured a 45 percent Public Works Administration grant in 1938. From that point forward, the replacement of the State Street Bridge and construction of a new subway tunnel under the river at State Street became inexorably linked.

Limited finances forced the project to be scaled back, however. The widening of the street was postponed, though the bridge was built to the original plans to accommodate street widening should funding for the project be made available within the estimated forty- to fifty-year life span of the bridge. Once the old bridge superstructure was removed, construction on the subway tunnel began with the removal of the huge concrete piers left in the riverbank from the old Scherzer bridge. Although the project

| Current (fifth) State Street Bridge | | | | | |
|---|---|---|---|---|---|
| Opened | Bridge type | Designed by | Constructed by | Cost | Status |
| May 28, 1949 | Chicago type, double-leaf bascule | Bureau of Engineering, Chicago Department of Public Works | Overland Construction Co. (superstructure), FitzSimons and Connell Co. (substructure) | $3.5 million | Currently in use |

Looking southwest as a commercial tugboat passes under the current State Street Bridge. © Kevin Keeley

was given priority status by the Department of Public Works, the bridge substructure was not completed until 1942, at which point construction was halted by order of the War Production Board. Shortages of steel and materials continued for several years after World War II, and it was not until October 1947 that work on the bridge resumed. It was finally completed eighteen months later. The bridge carried two continuous lines of fifty-ton streetcars on tracks to either side of the center truss until February 18, 1957, when service stopped and the tracks were later removed. In 1959 the State Street Bridge was converted to one-person operation and has remained in operation for more than sixty years.

The first State Street Bridge was an arch-truss superstructure swing bridge, 184 feet long and 35 feet wide with a wood-piling substructure and curb and fill approaches. Though the project was initiated in 1860 and given a City appropriation of fifteen thousand dollars, private interests postponed its construction. By 1862 the bridge was considered "a necessity" demanded by the increase in population and business, but local business-men still failed to agree on the details of the bridge.[1] Obtaining right-of-way added further delay, as the Galena & Chicago Union Railroad owned a strip of land between the north bank of the river and Wolcott Street (later renamed North State Street). At that time, only three bridges crossed the Main Channel, at Clark, Wells, and Rush streets. With the destruction of the Rush Street

State Street before 1871, looking north from Lake Street, with the first State Street Bridge in the background.

Bridge in November 1863, the need for a new bridge "became overwhelmingly apparent," and every effort was made to obtain the final subscriptions.[2] On May 30, 1864, an agreement was struck with the railroad giving the City title to the necessary land, as well as rights to construct a viaduct over its rail line. The first State Street Bridge opened to pedestrians in January and to wagon teams in June 1865.

Like all bridges built at that time, construction depended upon the support of local residents, businessmen, and land-owners to pay at least half the cost of the bridge. The first State

| First State Street Bridge | | | | | |
|---|---|---|---|---|---|
| Opened | Bridge type | Designed by | Constructed by | Cost | Status |
| June 6, 1865 | Swing, wood, hand operated | Fox & Howard | Fox & Howard | $36,000 | Destroyed by fire on Oct. 8–10, 1871 |

Street Bridge subscriptions totaled twenty-one thousand dollars, and City funds appropriated by the Common Council covered the rest. The State Street Bridge gave the city a ten-mile continuous street railway line, from Camp Douglas at 31st and Cottage Grove to Graceland Cemetery at Clark Street and Irving Park. The trolley line ran most of the length of Chicago through the Loop and ended two miles beyond the city's northernmost limit in the town of Lake View. In 1871 the line was severed when the wood State Street Bridge was destroyed in the Great Chicago Fire.

A second State Street swing bridge opened in December 1872, as the City lost little time in replacing the destroyed bridges. This iron bridge was the same length and a foot wider than its predecessor and was supported by a pile and a masonry center pier. Designed by the Keystone Bridge Company, it was referred to as a pivot bridge, but likely had a rim-bearing turntable and was offered in steam-engine or hand-powered models. City of Chicago Bridge Department records show that a hand-operated bridge was selected. It featured hydraulic-powered wedges at the end of the trusses meant to firmly support the bridge when closed. The wrought-iron turntable likely used an improved steel-cone center invented and patented by the company's general manager, John L. Piper. The *Piper center* bore more weight and could not be displaced; furthermore, the models' "cheapness, strength, and durability—render them greatly superior to any other turn-table offered to the public."[3]

Looking west down the Main Channel at the Rush Street, second State Street, and Clark Street Bridges in 1874.

In 1880 the Keystone Bridge Company replaced a wobbly turntable, foreshadowing future problems at State Street. The *Annual Report of the Department of Public Works* of 1885 reported that "the condition of many of the river bridges is by no means equal to the demand upon them. This is especially true of the

| Second State Street Bridge | | | | | |
| --- | --- | --- | --- | --- | --- |
| Opened | Bridge type | Designed by | Constructed by | Cost | Status |
| Dec. 1872 | Swing, iron, hand operated | Keystone Bridge Co. | Keystone Bridge Co. | $49,500 | Removed in May 1887 |

bridges at State, Twelfth, and Eighteenth streets . . . designed to meet the demand[s] of the past and . . . not therefore adapted to the wants of the present, either in strength or dimensions."[4] The second State Street Bridge operated for another year before being removed in May 1887 to make way for a new bridge.

The third State Street Bridge was a steel Pratt truss turned by a steam-driven, rim-bearing turntable. The bridge was designed by Abraham Gottlieb, who, after resigning as president of the Keystone Bridge Company of Pittsburgh, returned to Chicago in 1884 and began his own engineering firm. He became the first chief engineer for the World's Columbian Exposition of 1893, but resigned from the position a year later due to restrictions upon his authority. His firm, A. Gottlieb & Company, would design much of the ironwork for Chicago's Masonic Temple, built in 1892 at the northeast corner of State and Randolph. It was demolished in 1939, in part due to costly modifications to the foundation needed to accommodate construction of the State Street Subway.

This State Street Bridge was the same length and 3 feet wider than its predecessor, providing a 23-foot-wide roadway and two 7¼-foot-wide sidewalks. The foundation of the second bridge was repaired and reused. Constructed to meet an emergency, it was understood that in the near future, the bridge would be transferred to another location and replaced by a four-lane bridge; this, however, never occurred.

In 1897, as the bridge was being converted from steam to electric power, it was noted that the old limestone substructure of the center pier was badly disintegrated. Repair seemed impossible and a new structure inevitable. The bridge was about 5 inches out of level and could be swung only one way. When it was first built, the river depth was 12 feet; this was later dredged to 18 feet. The center pier had then settled 27 inches to one side, which went unnoticed with the bridge safely supported by the abutments all winter. In February 1898, City engineer John Ericson tested all the bridges in anticipation of the navigation season, which would begin in early April. When swung, the eleven-year-old State Street Bridge tipped to one side and threatened to fall into the river. Ericson declared that the bridge would be swung and left opened at the beginning of the navigation season unless repaired before April. This presented a potentially serious problem, as its closure would sever a major artery of the city's traffic and commerce.

Recent federal regulation of bridges and the new policy to remove center-pier swing bridges further complicated the situation, and a lack of City funds heightened it to a crisis. A new center pier would cost between thirty and forty thousand dollars—well

| Third State Street Bridge | | | | | |
|---|---|---|---|---|---|
| Opened | Bridge type | Designed by | Constructed by | Cost | Status |
| Sept. 7, 1887 | Swing, steel, steam powered | A. Gottlieb & Co. | A. Gottlieb & Co. | $24,400 | Closed on Oct. 8, 1901, and removed in 1902 |

Looking west in 1929 at the third State Street Bridge with the Dearborn Street Bridge and Clark Street Bridge in the background. Courtesy of David R. Phillips.

beyond existing City finances—and require a permit, which was not expected to receive federal approval. Though the ultimate would be a new bascule bridge, at this time bascule bridges were an unproven entity and financially untenable.

The City Engineering Department provided an unprecedented engineering solution, one that received some notoriety in the trade press. By repairing the existing center pier, the City could avoid the process of obtaining a federal permit. New fifty-foot pilings were driven around the center pier two feet from the old masonry. This circle of pilings was then secured with several wraps of steel cable. Next, the top six feet of masonry on the old pier were removed and used to fill the gap between the old and new piers. The repaired center pier was then capped

with an immense circular casting, supporting thirty-seven girders that radiated from the center like spokes on a wheel. Each thirty-inch-wide by fifteen-foot-long girder was securely bolted to the central casting that rested on the encircling piles to form a level surface. A new track was placed, and the bridge turntable was reinstalled. With this new arrangement, the encircling pier supported 75 percent of the weight of the bridge, allowing for its reopening and buying the City valuable time.

During this project, much of the protection pier was extracted to allow the pile drivers to work, and many old wood piles from the very first State Street Bridge were discovered. Identified by the telltale char marks from the Great Fire of 1871, they had been in place for more than thirty years.

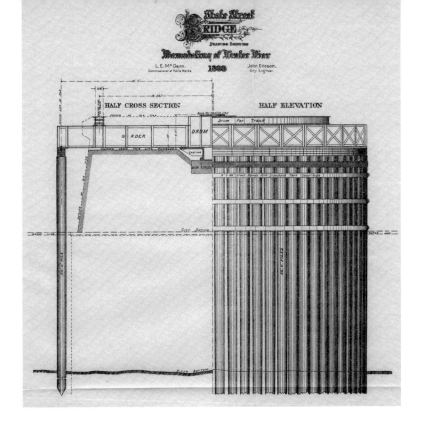

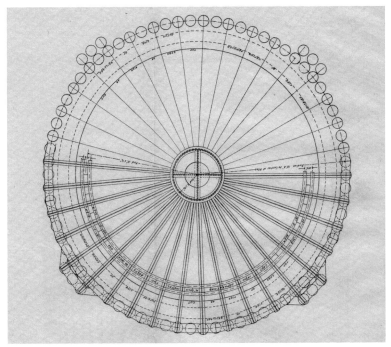

Design drawings of the improved center pier for the third State Street Bridge.

In 1900 this bridge was redecked, and the subfloor was preserved with two coats of red lead and oil paint. The inevitable demise of the span came in the fall of 1901, when it was finally closed to traffic. It was removed the following spring, and construction of a new bascule bridge proceeded.

The fourth State Street Bridge was completed in 1903 by the more suitably financed Sanitary District for the City of Chicago. This Scherzer rolling-lift bascule bridge was the largest highway bridge of its kind in Chicago and the eighth bascule built by the Sanitary District. It allowed a clear channel of 140 feet for navigation and increased flow of the river, which was needed for reversal of the Chicago River. The bridge operated using two 500-volt, 52-horsepower Westinghouse reversible street-railway motors and was powered by current from the trolley line crossing the bridge. Ownership and maintenance of this bridge were transferred to the City of Chicago upon its completion.

Maintenance of the bridge began around the turn of the twentieth century and lasted into the 1930s. In 1904 new street gates and electric warning bells, operated from the bridge houses, were installed at the approaches. Like most Scherzer rolling-lift bridges, this bridge also later required repeated replacement of rivets on the soleplates of the segmental girders and foundation track plates. With regular operation, rivets were sheared off of these plates, and in 1906, 1909, and 1911 countersunk bolts were used as a fix. The center and heel locks also required regular repair and adjustment. The bridge was repaved in 1907 and 1911, and in 1926 it was completely redecked. In 1937 two floor beams and the supporting gusset plates were replaced.

A 1930 report to secure financing for bridge repairs and replacement recommended that State Street receive a new bridge. City engineers argued that the existing bridge was outdated, de-

## Fourth State Street Bridge

| Opened | Bridge type | Designed by | Constructed by | Cost | Status |
|---|---|---|---|---|---|
| Feb. 28, 1903 | Scherzer rolling lift bascule | Scherzer Rolling Lift Bridge Co. | American Bridge Co. (superstructure), Lydon & Drews Co. (substructure) | $147,648 | Removed in May 1939 |

signed for the wagon and streetcar loads that prevailed around 1900. Furthermore, it was too narrow for street and trolley traffic and now needed to be twice as wide. The bridge also did not provide a sufficient channel width for shipping and navigation. The report noted that the rolling-lift design was economical and serviceable but insufficient to the requirements of the Chicago River and the revolution in traffic conditions over the previous twenty years. When this bridge was opened, the 138-foot draw did not align with a bend in the river, so the effective opening was only 110 feet and the existing structure caused an eddy in the current that frequently interfered with navigation. Furthermore, the bridge, which rested on pile foundations, had settled, causing both leaves to shift toward the center of the river. It was expected that the increase in traffic loads would accelerate this shift and exceed the City's ability to adjust the superstructure to accommodate such changes. Finally, the report stated that the planned subway could not "possibly be built under State Street while the present bridge, with its pile foundations, remains in commission."[5]

Federal approval for a new bridge at State Street would come with the condition that an unwanted 70-foot projection be removed from the north bank of the river. Additionally, building a new bridge with a proper bedrock foundation in coordination

Looking west in 1908 at the open fourth State Street Bridge as a propeller ship passes through.

with construction of the subway would offer significant savings in construction costs. In May 1939, the fourth State Street Bridge was removed in anticipation of a subway tunnel and a new bascule bridge at the site. Yet, to the repeated frustration of residents and commuters, this crossing would remain without a bridge for a full decade.

# DEARBORN STREET BRIDGES

LOCATION: 50 West, 307 North; Dearborn Street runs north and south and crosses the Main Channel 0.85 miles west of the river mouth at Lake Michigan.

HISTORICAL HIGHLIGHT: Dearborn Street was the location of Chicago's first moveable bridge, which was also the first bridge across the Main Channel of the Chicago River.

The fourth and current Dearborn Street Bridge, a postwar Chicago-type bascule, opened to traffic in 1963, two years behind schedule. The delay was caused by construction complications, arguments between the City and the contractors, a nationwide steel strike, and construction of the neighboring Marina City Towers. The most significant complication came with the construction of the caissons when 105-foot-deep shifting sand and silt kept builders from reaching the bedrock before beginning the foundation. Work building the caissons was further delayed by construction of the foundations for the Marina City Towers that was also under way. City engineers feared that excavation for the Dearborn Street Bridge and Marina City caissons at the same time would affect the foundations of nearby buildings and result in damage claims against the City. Construction on the bridge was halted until the Marina City caissons were completed.

The original Dearborn Street Bridge was the first drawbridge built across the Chicago River. This bridge was about 300 feet long and 16 feet wide, and it had a 60-foot draw. Soon after arriving in Chicago in November 1833, shipwright Nelson R. Norton began cutting lumber for this bridge on land near Michigan Avenue. Completed in August 1834, the Dearborn Street Bridge was very similar to a medieval drawbridge. An 1838 sketch from Common Council proceedings showed the addition of chain supports and installation of a block-and-tackle arrangement to operate the bridge. Ropes run through blocks were fixed at the ends of each leaf and to the top of each gallows frame, then run down to a wood drum at the approaches. The bridge tender would turn a wood drive gear, winching the ropes around the drum and raising each leaf of the bridge.

Local residents often remarked on the bridge's sinister cast, particularly when approaching its gallows frames at dusk. This drawbridge regularly required repairs from use and so became expensive and troublesome to operate. Most notably, it was once stuck open for more than forty-eight hours, and seemingly no amount of effort would close it.

After years of trouble, debate, and political wrangling, the Common Council finally approved the bridge's removal on July 8, 1839. South Side interests wanted to eliminate the bridge com-

## Current (fourth) Dearborn Street Bridge

| Opened | Bridge type | Designed by | Constructed by | Cost | Status |
|---|---|---|---|---|---|
| Oct. 23, 1963 | Chicago type, double-leaf bascule | Division of Bridges and Viaducts, Chicago Department of Public Works | Overland Construction Co. (superstructure), States Improvement Co. (substructure) | $6,832,000 | Currently in use |

Looking northwest at the current Dearborn Street Bridge in the fall of 2009. © Kevin Keeley

pletely to prevent the growing wagon trade from reaching North Side merchants. Citizens gathered before dawn the following day and tore the Dearborn Street Bridge down to ensure that the council could not reverse its decision, as had happened a year earlier. This ancient design would not be seen again in Chicago for more than fifty years; however, the bascule design would eventually prove to be the ideal bridging solution for the city.

This bridge allowed ships to traffic the river and connected

## First Dearborn Street Bridge

| Opened | Bridge type | Designed by | Constructed by | Cost | Status |
|--------|-------------|-------------|----------------|------|--------|
| Aug. 1834 | Gallows frame, wood, hand-operated bascule | Nelson R. Norton | Nelson R. Norton | Unknown | Torn down in 1839 |

Chicago's first drawbridge at Dearborn Street in the open position, looking east toward Lake Michigan. Revised by visualizedconcepts, inc.

the North and South sides in a manner far superior to a ferry. The need to cross the river remained, though, and within a month the State Street ferry relocated to Dearborn Street. After the navigation season ended, a scow bridge was substituted. It was an imperfect solution; as the *Chicago American Democrat* reported in February 1840, "The old scow bridge at the foot of Dearborn Street is now broken by the thaw," and residents had to step carefully when crossing between boards and scows.[1] The scow bridge was removed in the spring to allow for the passage of ships, and ferry service was resumed, which re-

mained at Dearborn Street for a long time. The uncertainty and inadequacy of a ferry instead of a bridge was well illustrated in June 1849 when a passenger named William Clark was thrown from the vessel, after a knot in the line suddenly checked its progress and hurled him into the river. Fortunately, Clark was an excellent swimmer and escaped with only the loss of his hat and umbrella. Regardless, Dearborn Street would not receive a second bridge for almost fifty years.

Dearborn Street received its second drawbridge in 1888. The superstructure of the old Wells Street Bridge, first constructed

Looking east in 1904 at the second Dearborn Street Bridge.
Courtesy of the State of Illinois Archives.

in 1872, was reused at Dearborn Street. It was floated by scow down the river and installed on a new foundation and center pier. This fantastic feat is recounted in two articles reprinted in the accompanying sidebar.

In 1893 the boiler for this bridge gave out and was replaced by the auxiliary boilers from the Rush Street Bridge. In 1897 the bridge was converted from steam to electric power, and in 1901 the Chicago Union Traction Company replanked the entire roadway. On two consecutive days in 1902, the bridge sustained collisions from ships. On May 22, the steamer *Lycoming* collided

with the north abutment of the bridge; the next day, while City bridge engineer Thomas Pihlfeldt was inspecting this damage, the steamer *Montana* hit the bridge's center-pier protection. The force of the collision momentarily lifted the bridge four feet up in the air and twisted the streetcar tracks. The ship swung around from the impact and hit the north abutment, causing tons of stone to tremble. Afterward, the bridge would not close properly; it was taken out of service, and streetcars were rerouted over the Clark Street Bridge. The collisions were the result of a tugboat drivers' strike happening at the time. Pihlfeldt reported that "the captain was to blame" and feared that, at this rate, the city might soon "not have any bridges left."[2] Though the bridge was soon repaired and put back into service, it was permanently removed in 1907 to make way for construction of a new bascule bridge.

The third Dearborn Street Bridge was a Scherzer rolling-lift span designed to improve navigation by eliminating the objectionable center pier of the old swing bridge. Constructed by the Sanitary District, this bridge was handed over to the City after its completion. The improved design did not eliminate collisions from ships, however. On September 18, 1911, the freight bark *Iron King* ran into and disabled the Dearborn Street Bridge after a tow line snapped. The heavy, three-masted ship rammed through the bridge pilings, bent the steel supporting girders almost in half,

## Second Dearborn Street Bridge

| Opened | Bridge type | Designed by | Constructed by | Cost | Status |
|---|---|---|---|---|---|
| 1888 | Swing, wood, hand operated | Fox & Howard | Fox & Howard (superstructure), FitzSimons and Connell Co. (substructure) | $22,820 | Removed in 1907 |

*Chicago Daily Tribune*, March 25, 1888, 10

## THE BRIDGE WILL BE MOVED TODAY

★ ★ ★

*Final Arrangements for Transporting Bridge Dearborn Street Completed Yesterday*

The Wells Street Bridge is to be moved this morning—at least so the Keystone Bridge Company says—and appearances yesterday seemed to bear out the statement. Men were at work on the scows that are to be used, and by night everything was in readiness for the undertaking. Two scows were fastened together by heavy beams, and on this double arrangement were erected huge horses made of the heaviest timbers. These are riveted together and strengthened in every possible way, and make a structure that will bear almost any weight. Another of these particular-looking crafts was constructed on the other side of the river.

As they stood last evening the tops of the horses were some distance above the bridge and of course the first thing to be done will be to get them low enough to float under the ends of the bridge. This will be done by pumping water onto the scows. When they have settled far enough they will be floated under the bridge and the water pumped out. As they rise they will lift the bridge from its present resting-place. How great the buoyancy of the scows will have to be can be realized when it is known that the bridge is estimated to weigh ninety-seven tons.

Considerable difficulty will doubtless be experienced in getting the bridge through Clark street draw. The space there is only just about great enough to admit of the scows passing through. The danger of capsize to which THE TRIBUNE called attention, appears to the ordinary observer to be great, but the contractors do not seem to fear anything of the sort. The combination scows are but little wider than the bridge itself, and when it comes to holding ninety-seven tons so far above the surface of the river it would seem as though the structure would be a bit top-heavy.

The turntable will be taken out just as soon as the bridge is away, and will be sent down on another scow and put in place at Dearborn street before the bridge is put there.

*Chicago Daily Tribune*, March 27, 1888, 1

## THE CRUISE OF A BRIDGE

THE WELLS STREET STRUCTURE VOYAGE TO DEARBORN STREET

*A Trip Made in Safety and Witnessed by Thousands of Persons—Engineer Strobel*
*Issued His Commands from "the Bridge"—A Description of the Journey—An Accident*
*Narrowly Averted at Clark Street—The Destination Reached at 6:45 O'Clock.*

There is no bridge at Wells street, yet there is still a Wells street bridge; there is no Dearborn street bridge, though there is a bridge at Dearborn street. Three thousand persons, male and female, shivered two mortal hours in the March wind of yesterday afternoon in order to spring this riddle on their friends before the answer should appear in the morning papers. And this is the answer: Though there is still a Wells street bridge there is no bridge at Wells street because the Wells street bridge has been floated down the river; there is thus a bridge at Dearborn street, to be sure, but the Wells street bridge at Dearborn will never be the Dearborn street bridge until it is possible to cross it without the aid of a derrick or a balloon.

After many days of promises and of getting up at unholy hours in the morning the multitudes who had set their hearts on seeing Wells street bridge sail majestically down to Dearborn street or plunge ignobly to the bottom of the creek at 5:30 yesterday gave vent to their feelings in a prolonged shout of exultation. At that hour the bridge moved slowly off the center pier. Notwithstanding Engineer C. L. Strobel of the Keystone Bridge Company, attired in a new lavender spring overcoat and red walking gloves with heavy black stitching on the backs, stood solitary and calm on the top of the swaying structure—notwithstanding this sublime spectacle of confidence based upon infallible scientific calculations, the crowd was not unmindful of the duty which always derives upon onlookers under such circumstances. Their shouts of warning and command were incessant, but the men engaged in the work paid not the slightest attention.

The rain and sleet of Sunday retarded preparations, so that the moving was accomplished twelve hours later than intended. At 10 o'clock yesterday morning the work was abandoned for the same reason. An hour or so of warm rain and south wind however, decided the men to go on with the work. The two pairs of scows with their deck-works of heavy pine timbers were soon placed under the bridge on each side of the center pier, the scows having been filled with water till their decks were nearly on level with the surface of the river. The weight of the bridge, denied of all flooring and other woodwork, was about ninety tons. In order to raise the bridge off the center pier it was therefore necessary to pump that much water out of the scows. The pumping began at noon and continued until the voyage began. Sixteen men worked four hand-pumps on each scow. When it was observed that a portion of the weight of the bridge was borne by the scows the pumping out process was materially hastened by stream-siphons operated from two tugs—the Alert and the Allen. When the bridge began to clear the pier a force of men went to work taking off the rims of iron which serve for bearings for the wheels upon which the bridge revolves. These rims, the wheels, and the heavy iron cap of the pier, with its track and cog-rim, were left behind. They will have to be transferred to the same position at Dearborn street, however, before the bridge can be lowered and the scows taken from under it.

The structure swayed from side to side and the scows seemed hardly able to bear their burden.

READY TO BEGIN THE VOYAGE

"Strobel knows what he's about," said a man in a tarpaulin. "You see there are two scows eight or ten feet apart upon which the timbers supporting the bridge are placed. The bridge will make its voyage end foremost. If there were only one scow at each end of the bridge a little inclination would ship the water in it to one side and over the thing would go. There being two scows this is impossible. While the bridge is being turned endwise with the current you observe that the greatest care is used to prevent any sudden jar that might change the centre of gravity."

With the tug Alert behind and the Allen in front to guide the strange craft, and with Mr. Strobel, dapper and natty, walking aloft issuing orders, the bridge was floated down to Clark street. As a precaution against accidents a guy-line made fast to schooners along the south shore was paid out by a gang of men on the rear larboard corner of the hindmost scow.

When the Allen blew her whistle Clark street bridge could not have held a dozen more people. An ineffectual attempt was made to drive them off, and then the bridge turned. It was opened barely fifteen minutes. The distance between the piers at Clark and Dearborn streets is so short that it was necessary to begin to turn the front end of the floating bridge to the north before the rear scows had cleared the draw. What might have been a serious accident—the only one of the trip—was narrowly escaped here. While the Allen was pulling at the front scows those behind drew dangerously near the pier. A workman saw the danger and yelled to the Captain of the Alert:

"Back 'er, Cap, quick!"

Then he and half a dozen of his fellows threw their shoulders against the pier and prevented a collision, though the end of the scow grated against the heavy timbers with a thrillingly suggestive sound.

At 6:45 the Wells street bridge rested on the scows above the pier at Dearborn street.

twisted the shafts in the operating machinery, and hit the concrete foundation before coming to rest with its pilot house lodged in the middle of what remained of the sidewalk. Sixteen tons of structural and machine steel had to be replaced, at a cost of $7,696.78 (or about $178,000 in today's currency), which was charged to the ship's owners. Traffic was restored on September 22.

By the 1950s, the roadways of the Dearborn Street Bridge had become too narrow to accommodate the burgeoning automobile traffic. The dramatic increase in postwar automobiles, however, was not the only contributor to the obsolescence of this bridge. The advancement of bridge technology now allowed for operation of a new bascule bridge by a single bridge tender, while the existing bridge required at least two. In November 1959, the old Scherzer rolling-lift bridge was closed to traffic and removed in favor of a new Chicago-type bascule bridge that featured one-person operation.

Looking east as a streetcar crosses the third Dearborn Street Bridge in 1908.

## Third Dearborn Street Bridge

| Opened | Bridge type | Designed by | Constructed by | Cost | Status |
|---|---|---|---|---|---|
| 1907 | Scherzer rolling lift bascule | Scherzer Rolling Lift Bridge Co. | Geo. W. Jackson, Inc. (superstructure), Great Lakes Dredge & Dock Co. (substructure) | $316,067.72 | Removed in Nov. 1959 |

# CLARK STREET BRIDGES

LOCATION: 200 West, 307 North; Clark Street runs north and south and crosses the Main Channel 1.2 miles west of the river mouth at Lake Michigan.

HISTORICAL HIGHLIGHT: During 1839 and 1840, construction of the very first bridge at Clark Street was the center of a divisive political battle that pitted North Side and South Side merchants against one another. Since then, Clark, Wells, and Randolph streets have received the greatest number of bridges (the current bridge is the eighth at each location).

The current Clark Street Bridge was completed in 1929, less than three months after the *Sandmaster* destroyed the previous bridge. Its grand-opening parade on July 10 included several members of the Sac and Fox tribes in full regalia, as well as displays tracking the city's progress from the days when Clark Street was a simple Indian trail. This second-generation Chicago-type bascule is a fixed-trunnion double-leaf drawbridge. It measures 245 feet from trunnion to trunnion and provides a clear span of 215 feet. The riveted steel-pony truss is 72 feet wide and carries four lanes of traffic and two sidewalks cantilevered off of the outside of the two bridge trusses.

The two Beaux Arts–style bridge houses on either corner of the bridge feature mansard tin roofs with a raised diamond pattern. First popularized by French architect François Mansart in the seventeenth century, this style of roof is four sided, with two slope angles to each side, and constructed in such a way that the lower slope is steeper than the upper. These identical bridge houses with matching approaches are constructed from reinforced concrete with a scored concrete veneer. The Clark Street Bridge was converted to one-person operation in 1956 and remains in use after more than eighty years of service.

Construction of the first Clark Street Bridge in July 1840 was the subject of a year-long political battle between the North and South sides. Early Chicago politics were contentious, occasionally pitting whole sections of the city against one another while personal fortunes hung in the balance. After the removal of the Dearborn Street Bridge, the city was reduced to two bridges, at Kinzie Street over the North Branch and over the South Branch above Randolph Street. Although a new bridge was proposed at Clark Street, Alfred T. Andreas's *History of Chicago,* published in 1884, best described the problem:

*Every night there came up out of the south a great fleet of prairie schooners that anchored on the Reservation.\* It often*

| Current (eighth) Clark Street Bridge | | | | | |
|---|---|---|---|---|---|
| Opened | Bridge type | Designed by | Constructed by | Cost | Status |
| July 10, 1929 | Chicago type, double-leaf bascule, steel construction, electric powered | City engineer Loran Gayton and city bridge engineer Paul Shioler | FitzSimons and Connell Co. (substructure), Ketler-Elliott Co. (superstructure) | $1,529,492.77 | Currently in use |

Looking west at the open current Clark Street Bridge with the open LaSalle Street Bridge and half-open Wells Street Bridge in the background in the spring of 2010. © Kevin Keeley

## First Clark Street Bridge

| Opened | Bridge type | Designed by | Constructed by | Cost | Status |
|---|---|---|---|---|---|
| July 1840 | Pontoon float swing, wood, hand operated | Unknown | William B. Ogden | $3,000 | Destroyed by flood on Mar. 12, 1849 |

*numbered five hundred, and came laden with wheat and corn and all sorts of produce. All the warehouses were in that day built on the north bank of the river.\*\* The South Side opposed the Clark-street bridge, in order that their prairie schooners might not reach those warehouses, and thus be compelled to trade on the south bank. The old Dearborn bridge, the first drawbridge ever built in the city, had been demolished in 1839, and a scow ferry substituted. At Clark Street, there was another ferry; these were not of the most approved pattern. They were simply scows hauled to and fro by ropes. The North Side warehouses were in sore distress. They needed a connection with the other two towns. The [City] Council was evenly divided. At the time when the question was at its height, Messrs. Newberry and Ogden presented to the Catholic ecclesiastical authorities the two*

*blocks now occupied by the [Holy Name] cathedral. It was said at the time that the present was to influence votes on the bridge question. It undoubtedly was. The North Side won her bridge. Mayor Raymond cast the deciding vote.[1]*

Additional political maneuvering followed this agreement, with efforts made to put the new bridge at Wells Street instead of Clark. This failed, and Clark Street received a pontoon-float swing bridge thanks to William B. Ogden, the first of its kind in the West. This bridge ensured development of the street into a major thoroughfare connecting the North and South sides. The Common Council approved a bridge at Wells Street a year later financed by citizen subscriptions.

The Clark Street Bridge and four others were destroyed in the flood of 1849, as recounted in "The Flood of 1849" sidebar in the introduction.

A second, more substantial, Clark Street pontoon-float swing bridge replaced the first that was destroyed in the flood of 1849. It was still operated manually, but a geared turntable under one end assisted the bridge tender in opening and closing the heavier, more substantial floating pier. The bridge had the added advantage of sitting ten to twelve feet above the water, allowing smaller boats to pass underneath. The City paid fifteen

---

*The Reservation refers to Fort Dearborn and a large tract of federal land south of the Chicago River that stretched from the fort east to the lakeshore. It included most of the high ground along the south bank of the Main Channel of the river.

**South Siders had no warehouses—the whole south bank of the river was a miry swamp—except on the eastern side, which belonged to the U.S. military.

## Second Clark Street Bridge

| Opened | Bridge type | Designed by | Constructed by | Cost | Status |
|---|---|---|---|---|---|
| July 3, 1849 | Pontoon turntable swing, wood, hand operated | John Censor | John Censor | $2,800 | Wrecked by steamer on Sept. 8, 1853 |

hundred dollars of the total twenty-eight-hundred-dollar cost for the bridge, and the surrounding property owners provided the balance. The new bridge was completed within six weeks.

In 1853 the bridge was almost completely destroyed after the steamer *London* collided with it. The "sunken wreck" of a bridge blocked the river for several hours until it was dragged to one side and pulled up onto the north shore.[2] The remains were later towed away, repaired, and used at Polk Street. This was Chicago's first bridge to be fatally wounded by a ship. Accidents like this would continue and even become quite common through the end of the nineteenth century, though they did not usually result in the complete destruction of a bridge.

The third Clark Street Bridge was designed and built by City superintendent of public works Derastus Harper. It was a pivot bridge with sidewalks and a double carriageway built to carry both north- and southbound traffic. It improved on the Lake Street design built two years earlier, also by Harper. This bridge was 330 feet long and 30 feet wide. It was also the first bridge to incorporate a protection pier, which would soon be standard on every swing bridge in Chicago. Protection piers comprised a series of pilings outlining the open bridge. Situated in the middle of the river channel, these piers were designed to protect the open bridge from passing ships. The third Clark Street Bridge was considerably longer than any other bridge on the river, offering a 75-foot draw to ships on either side.

During the Beer-Lager Riots of 1855, this bridge was opened to delay an angry crowd and give police time to gather and prevent storming of the jail. The riots arose because of the enforcement by Mayor Levi Boone's administration of an ordinance prohibiting the sale of beer on Sundays and a sixfold increase of

## Third Clark Street Bridge

| Opened | Bridge type | Designed by | Constructed by | Cost | Status |
|---|---|---|---|---|---|
| July 4, 1854 | Pivot, wood, hand operated | Derastus Harper, city superintendent of public works | Derastus Harper, city superintendent of public works | $12,000 | Broke in two on July 28, 1858 |

Chicago's liquor-license fees. Then several tavern owners were arrested in violation of the ordinance and sentenced on Saturday, April 21; when a crowd gathered around the City courthouse, Mayor Boone ordered police to clear it. This resulted in nine more arrests. The next day, a mob of mostly German and Irish immigrants, intent on freeing the prisoners, gathered. After gaining control of the Clark Street Bridge, they crossed the river and confronted the mustered police force, and in the resulting clash one man was killed and several others were seriously wounded. It was Chicago's first civil disturbance. The next year, a strong German and Irish vote ousted Mayor Boone and his Know-Nothing Party, restoring the status quo and ending the disinterested, nonpartisan politics of early Chicago.

The third Clark Street Bridge also experienced serious structural problems. Its great length was too much for its design and materials, and a year after it was built, the top chords of the truss gave way and threatened to break in half. The bridge was repaired by binding the chords together with "an immense cable," and then later rerepaired by bolting the top chords together with heavy iron plates.[3] In spite of these fixes, on July 28, 1858, the bridge utterly failed. The structure broke in two and fell into the river, severing "the largest artery of communication in the city."[4]

The fourth Clark Street Bridge was more substantial and was built quickly to replace the collapsed bridge. This Howe-truss

Newspaper drawing, looking east in 1857, of the third Clark Street Bridge.

bridge was designed by Newton Chapin, a prominent bridge builder and contractor before the Great Fire.

Chronicled as "frisky under spring influences," in March 1859 this bridge was left unfastened, and a high wind suddenly blew it open, much to the surprise of the teams and pedestrians crossing at the time.[5] No harm was done, and the bridge was soon brought back into place.

This bridge was the busiest river crossing at that time and had to be replanked as early as November 1859. In January 1866, the City superintendent reported to the Common Council that the fourth Clark Street Bridge was "quite rotten and no longer

## Fourth Clark Street Bridge

| Opened | Bridge type | Designed by | Constructed by | Cost | Status |
|--------|-------------|-------------|----------------|------|--------|
| 1858 | Swing, wood, hand operated | Chapin & Co. | Chapin & Co. | $14,200 | Removed on Apr. 6, 1866 |

Photograph of the fourth Clark Street Bridge in the early 1860s. Courtesy of the Chicago History Museum.

safe for heavy travel."[6] It was removed that spring to make way for a new bridge.

The fifth Clark Street Bridge was the first combination wood and iron bridge in Chicago. This hand-operated bridge was 180 feet long and 32 feet wide. It incorporated the improved Howe truss and turntable patented by James K. Thompson. The advantages of this bridge were never fully realized, however, because it was destroyed five years later in the Great Fire.

Like a phoenix, the city rose from the ashes of the Great Fire, and a sixth Clark Street Bridge opened to traffic in 1872.

This Howe-truss bridge was 5½ feet wider than, and the same length as, its predecessor. This Howe-truss bridge served Clark Street for seventeen years and was not replaced until 1889, at which point it was refurbished and installed at Webster Street, where it served capably for another fifteen years.

The North Chicago Street Railroad paid for the seventh Clark and the seventh Wells Street Bridge superstructures and ongoing maintenance costs. In covering these costs, the streetcar company gained important connections to the Loop and circumvented the twenty-five-thousand-dollar annual fee charged for

The fifth Clark Street Bridge before the Great Fire of 1871.
Courtesy of the Chicago Public Library, Archives and Special Collections.

The remains of the Clark Street Bridge after the Great Fire of 1871.
Courtesy of the Chicago Public Library, Archives and Special Collections.

## Fifth, sixth, and seventh Clark Street Bridges

| Opened | Bridge type | Designed by | Constructed by | Cost | Status |
|---|---|---|---|---|---|
| June 9, 1866 | Swing, wood and iron, hand operated | Board of Public Works and James K. Thompson | Thomas Mackin | $13,800 | Destroyed by fire on Oct. 8–10, 1871 |
| Jan. 9, 1872 | Swing, wood and iron, hand operated | Fox & Howard | Fox & Howard | $32,000 | Moved to Webster Street in 1889 |
| 1889 | Swing, steel, steam, then electric powered | North Chicago Street Railroad | Variety Iron Works (superstructure), FitzSimons and Connell Co. (substructure) | $186,562 | Destroyed by collision on Apr. 30, 1929 |

use of the La Salle Street Tunnel. The City paid for construction of both bridge substructures, but North Chicago Street Railroad designed the bridges, specifying four-track roadways that could carry a streetcar line and a lane for vehicles in both directions. In 1889 the *Annual Report of Public Works* praised the steel Pratt-truss Clark Street Bridge as "first-class in every particular."[7]

Swinging open more than ten thousand times a year, after five years the iron rollers of the seventh Clark Street Bridge were worn down to the point that the pinion bracket was resting on the circular rack surrounding the turntable. The most economical solution was chosen, and three-eighths of an inch was removed from the top of the circular rack; this was made possible by the fact that the rollers had worn uniformly. In 1896 assistant engineer Jules E. Roemheld of the City Bridge Department invented an ingenious method for replacing turntable rollers. Using an inter-locking wedge system, Roemheld's method raised the bridge two and a half inches on its own power to allow replacement of the old iron rollers with a new set of cast-steel rollers in just four hours. This saved the City a great deal of time and money, as the old method of swapping rollers required taking the bridge out of service, employing a large workforce, building a heavy timber falsework on the center pier, and using hundreds of jackscrews to raise the bridge.

In 1897 the bridge was converted from steam to electric power, and in 1907 the bridge roadway was repaved with A. F. Shuman's composite wood and asphalt pavement between the outer streetcar rail and the wheel guards. By 1926 the Clark Street Bridge was regularly carrying the heaviest streetcar traffic of any city bridge. The eyebars at the joints of the superstructure had to be either reinforced or replaced and new cast-steel rollers installed in the turntable. The worn surface and underlying four-

Looking east downriver from LaSalle Street at the open (sixth) Clark Street Bridge in the 1880s. Courtesy of Walter Lewis and the Maritime History of the Great Lakes.

inch block roadway were removed, and the entire bridge received asphalt paving.

On April 30, 1929, with an eighth Clark Street Bridge under construction, this old swing bridge met a premature end after being hit by a sand barge. The *Sandmaster,* a chronic Chicago bridge molester, approached at too rapid a pace and rammed the bridge before it had time to fully open. The steel prow of the ship carved into the end of the bridge. Unable to sustain the terrific impact, the old bridge was knocked from its turntable rollers and came to rest tilting at a dangerous angle over the river. The impact from the *Sandmaster* moved the six-hundred-ton Clark Street Bridge a total of seven feet; one foot farther, and the bridge

The seventh Clark Street Bridge, looking east in 1926, with the State Street Bridge in the background. Courtesy of the State of Illinois Archives.

Aerial view looking east at the *Sandmaster* after hitting the seventh Clark Street Bridge in 1929. Courtesy of David R. Phillips.

would have been driven into the river. It was decided to demolish the bridge and rush completion of the new bridge rather than make repairs. An emergency order was given to the Great Lakes Dredge & Dock Company to remove those portions of the bridge that obstructed navigation. Within a day, enough of the bridge superstructure had been removed for the sand boats to pass, and a day later the south draw was made clear to navigation.

Subsequent investigation by the City found that the 251-foot *Sandmaster* had rammed or damaged thirteen Chicago bridges a total of forty-four times in the previous three years. In cases like this, when fault was proven, the City's Corporation Counsel pursued restitution from the ship's owners. The Ketler-Elliott Company, as part of their construction contract, would later remove what remained of the seventh Clark Street Bridge.

# LA SALLE STREET BRIDGE AND TUNNEL

LOCATION: 100 West, 307 North; La Salle Street runs north and south, crossing the Main Channel one mile west of the river mouth at Lake Michigan.

HISTORICAL HIGHLIGHT: La Salle Street's first connection across the river was not a bridge but a tunnel that was used from 1871 to 1939.

The current La Salle Street Bridge, the first at this location, is one of the most ornamental bridges in the city. It received special treatment after the widening of La Salle Street between Washington Street and Lincoln Park, a key element of the *Plan of Chicago*. This bridge was designed as an important connection over the river in the plan to rationalize city traffic patterns. Once completed, La Salle Street provided a major north-south artery consistent with Daniel Burnham and Edward Bennett's plan. The ornamentation and four bridge towers reflect the designers' attention to aesthetics, as two of the towers are entirely superfluous. The Baroque-style ornamentation on the four bridge houses, influenced by Bennett, features mansard roofs, swags, and rusticated masonry very similar to that used in the pylons situated on either side of the Balbo Street entrance to Grant Park.

Yet even with such treatments, advocates of the City Beautiful movement were dissatisfied with the bridge and its neighbor at Clark Street. Eugene Taylor, manager of the Chicago Plan Commission, was quoted in 1930 as saying of the bridges, "They look like the devil."[1] These structural twins, which both had huge steel trusses that protruded above street level, were compared to billboards hiding pleasing scenery. A City engineer of the day explained that the bridges were constructed with the trunnions above the street level in order to meet federal river-clearance requirements. Thus, much of the superstructure was placed above the road deck, particularly at the approaches. The effect was later softened by adding ornamental iron railings to the sidewalks.

The very first connection, a tunnel, between the North and South sides at La Salle Street was started on November 3, 1869. The City-owned La Salle Street Tunnel opened two years later and was modeled after the Washington Street Tunnel (completed two years earlier) that connected the South and West sides. Originally designed for pedestrian and horse-drawn vehicles, both tunnels provided important escape routes during the Great Chicago Fire. They were developed to provide a constant con-

| LaSalle Street Bridge | | | | | |
|---|---|---|---|---|---|
| Opened | Bridge type | Designed by | Constructed by | Cost | Status |
| Dec. 20, 1928 | Chicago type, double-leaf bascule | City engineers Donald Becker, Thomas Pihlfeldt, and Clarence Rowe | Strobel Steel Construction Corp. (superstructure), Central Dredging Co. (substructure) | $1.6 million | Currently in use |

A view of the LaSalle Street Bridge, looking southwest in the winter of 2009. © Kevin Keeley

nection for traffic as an alternative to the swing bridges that were so often opened for river traffic. Construction of a tunnel was suggested as early as 1844 as a more reliable and continuous means for crossing the river.

However, the dark, dank tunnels did not attract as much traffic as expected. Underutilized and falling into disrepair, the tunnels became more valuable to the streetcar companies than to horse-drawn vehicles and pedestrian traffic in the 1880s. The

## LaSalle Street Tunnel

| Opened | Tunnel type | Designed by | Constructed by | Cost | Status |
|---|---|---|---|---|---|
| July 4, 1871 | River tunnel | William Bryson | Robert E. Moss | $566,000 | Closed in Nov. 1939 |

Grant Park pylons designed by architect Edward Bennett at Balbo Street. © Patrick McBriarty

two-thousand-foot-long La Salle Street Tunnel was seen as a reliable means of carrying commuters in and out of the Loop under the river and thereby avoiding the unpredictable swing bridges. The reversal of the river in 1900 affected the tunnels, however; the Sanitary District's planning for a minimum river depth twenty-six feet from the existing seventeen feet necessitated the removal or lowering of the tunnels under the river.

The City placed the burden for this on the streetcar companies using the tunnel. A novel engineering solution was devised through which the old tunnel was removed, and a deep trench was dredged from the riverbed to allow for installation of a new tunnel and a river depth of at least twenty-seven feet. To accomplish this, a series of sixty-three-foot steel I-beams were driven along each bank to contain the river, and open trenches were dug on the opposite side for construction of the tunnel approaches. At a shipyard on Goose Island, two watertight steel-plate tubes measuring twenty-four feet wide by forty-one feet high were erected, and their ends were capped and sealed. The air trapped inside the tubes provided buoyancy so they could be towed up the river, maneuvered, and then sunk into place. They mated

Decorative railing on the LaSalle Street Bridge. © Kevin Keeley

Passage through the LaSalle Street Tunnel by streetcar.

with the I-beams along the riverbank and were then covered with fill in the riverbed. The water inside the tubes was pumped out, and openings were cut so the approaches could be connected. The new tunnels opened to streetcar traffic on June 22, 1912; by 1939, however, use of the La Salle Street Tunnel had greatly diminished, and, following the construction of the Dearborn subway station, the tunnel was closed and never reopened.

The fabricated steel tubes afloat on the river during tunnel reconstruction in November 1910. Courtesy of the Chicago History Museum.

# WELLS STREET BRIDGES

LOCATION: 200 West, 309 North; Wells Street runs north and south and crosses the Main Channel 1.4 miles west of the river mouth at Lake Michigan.

HISTORICAL HIGHLIGHT: The current Wells Street Bridge is one of four double-deck bascule bridges. As with the Clark and Randolph Street bridges, this is the eighth at this location.

This bridge is a fixed-trunnion, double-leaf, double-deck bascule bridge measuring 268 feet from trunnion to trunnion and offering a 231-foot clear channel. It is longer than both of the Michigan Avenue and Lake Street double-deck Chicago-type drawbridges that preceded it. The bridge carries double tracks for use by the Chicago Transit Authority's (CTA) elevated trains on the top deck. The lower deck has double sidewalks cantilevered off of the outside of the bridge truss, and the roadway is 72 feet wide. This thru-truss superstructure was constructed using steel and riveted gusset-plate connections fabricated by the Fort Pitt Bridge Company.

The eighth Wells Street Bridge was the second double-deck bridge built at this location. As in the case of its Lake Street counterpart, an earlier swing bridge had been modified to carry elevated trains on its upper deck. The construction of the current double-leaf bascule bridge was an amazing feat of engineering and coordinated effort. Maintaining elevated-train traffic throughout its construction, the trains were interrupted for only a single weekend while the bridge was being lowered for the first time. A key connection for daily commuters in and out of the Loop, the new bascule bridge was built in the upright (or open) position to retain use of the existing double-deck swing bridge. This allowed "L" traffic to continue uninterrupted throughout 1921, until 7:00 p.m. on the first Friday in December. Traffic was then halted, and the old bridge was swung open. The "L" tracks were pulled up, and the old bridge was cut down to allow the leaves of the new bridge to be lowered into place. The upper bridge deck was completed, rail approaches and tracks were installed, and "L" traffic was restored later that same weekend. Work crews labored around the clock to ensure a normal Monday-morning commute.

The rest of the bridge construction was completed three months later, and street traffic resumed. This tremendous success is mostly attributed to City engineer of bridges Thomas Pihlfeldt. He had orchestrated a similar replacement of the Lake Street Bridge six years earlier, interrupting elevated traffic for just a week and attracting the attention of engineering journals. It was reported that "the two bridges held a place in Pihlfeldt's memory as his proudest accomplishments in the city's service."[1]

| Current (eighth) Wells Street Bridge | | | | | |
|---|---|---|---|---|---|
| Opened | Bridge type | Designed by | Constructed by | Cost | Status |
| Feb. 11, 1922 | Chicago type, double-deck, double-leaf bascule | Bureau of Engineering, Chicago Department of Public Works | FitzSimons and Connell Co. (substructure), Ketler-Elliott Co. (superstructure) | $1,341,925 | Currently in use |

Looking west at the current Wells Street Bridge and massive Merchandise Mart building to the right in the background in 2009. © Kevin Keeley

The size and weight of this bridge presented some engineering challenges beyond its double-deck design. The additional weight of the double-deck structure required a new substructure for the counterweight pit in order to support the bridge. Piers reaching down into the bedrock supported the new counterweight pit, which was reinforced with columns and cross-girders to provide added support to the trunnion bearings. On the lower deck, the break in the roadway for opening the bridge was located between the trunnions, where most bascule bridges feature the break outside of the trunnions to better support street traffic and reduce dynamic load from vehicles moving on and off the bridge leaf. The resulting vibration from such loading can give the bridge a tendency to bounce or, even potentially, open unexpectedly. This problem was solved by adding heel locks designed specifically for the Lake Street, Michigan Avenue, and Wells Street bridges. The heel locks effectively provide additional

Construction of the south-bank substructure in 1920 for a new Wells Street Bridge, showing the old double-deck swing bridge in the background. Courtesy of CDOT.

support to the bridge at the approaches. The gear train in the operating machinery was also made from high-strength steel to better accommodate the dynamic loads of the heavier bridge.

In November 1935, this Wells Street Bridge was rammed by the boat *H. Dahlke.* Captain Sherman Barnard, a former tugboat captain in his second day in command of the *Dahlke,* was operating in thick fog. Expecting the bridge to open on demand, the ship slammed into the side of the unyielding Wells Street Bridge.

Both captain and mate barely escaped to safety before the ship's pilothouse was utterly destroyed. The heavy boat continued forward, and a steel derrick behind the pilothouse cut into the bridge another fifteen feet, tearing away the east sidewalk and bending back portions of the bridge superstructure. Responsibility for the accident was disputed, and the estimated damage to the bridge totaled six thousand dollars (approximately ninety-three thousand in 2010 dollars). The following year, Wells Street Bridge suffered three malfunctions within ten days of one another; these were attributed to the ramming by the *Dahlke,* and additional repairs were made. In 1936 new creosoted block pavement was installed, and the four main columns of the approaches were reinforced with horizontal trusses. In 1956 the old bridge deck was removed and replaced with an open-grid steel roadway.

In 1977 a driver was injured and her vehicle destroyed while attempting to cross the bridge. Raphan Boonying drove across the bridge's lower deck but was trapped on the other side by the warning gate with her car straddling the roadway and the bridge deck. As the bridge began to open and the car started sliding back toward the river, the upper level pinned the car to the roadway and began crushing it. The bridge tender, finally hearing Boonying's screams, stopped operation of the bridge to investigate. Boonying was extricated from the wreckage and taken to the hospital, where she later recovered. The Wells Street Bridge has operated without incident ever since.

Closed to street traffic in November 2012, a complete rehabilitation of this 90-year-old bridge was initiated. CTA traffic on the upper deck was maintained except during two 9-day shutdowns (in April for the south leaf and May for the north leaf), allowing replacement of the two truss sections over the river. Amazingly these half-million pound truss replicas were fabricated, floated by barge, and attached to the refurbished approach structures. Never before has the City of Chicago un-

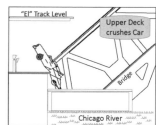

Sequence of events that led to crushing of a car by the double-deck Wells Street Bridge. © Patrick McBriarty

dertaken such a complex and intricate bridge project on this scale. Preserving the architectural elements of the bridge, work was completed at the end of 2013. This landmark undertaking is captured in the documentary *Chicago Drawbridges*.

The first Wells Street Bridge was built in 1841. At the time, it was one of five bridges crossing the Chicago River and one of only two bridges across the Main Channel. It would be sixteen years before a third bridge, built at Rush Street, would cross the Main Channel. Today, ten bridges cross the Main Channel.

This first Wells Street Bridge was a pontoon-float swing bridge. It had wood approaches supported by pilings at either bank, with a floating draw in the middle. John Van Osdel, who later opened Chicago's first architectural firm, designed and built the bridge. Early float bridges such as this one were of all-wood construction and had a life expectancy of only about five years. The first Wells Street Bridge was replaced after six years of heavy use.

## First Wells Street Bridge

| Opened | Bridge type | Designed by | Constructed by | Cost | Status |
|--------|-------------|-------------|----------------|------|--------|
| 1841 | Pontoon swing, wood, hand operated | John Van Osdel | John Van Osdel | $3,500 | Removed in 1847 |

The second Wells Street Bridge was a pontoon swing design with improved floats fabricated from boiler iron. Completed in 1847, it was destroyed two years later by the flood of March 12, 1849, recounted in "The Flood of 1849" sidebar in the introduction. The iron floats survived and were salvaged.

The third Wells Street Bridge was a new pontoon-float swing design operated by a turntable supported by pilings under one end and floats on the other. It was built in 1849 and likely reused the recovered floats from the second bridge. The third Wells Street Bridge was removed in 1856 to make way for a new pivot bridge.

The fourth bridge was an early pivot bridge, designed and built by Derastus Harper, that was similar to the design built at Clark Street in 1854. This 190-foot bridge differed, however, in that it had three instead of Clark Street's two arch trusses and was of much stronger and heavier construction.

At the end of 1861, the fourth Wells Street Bridge showed signs of weakness and was strengthened using a cable to relieve the strain on the chords. A thorough examination of the bridge that August, however, showed the woodwork to be significantly decayed and the trusses weak and likely to fail at anytime. Replacement was recommended. During the dismantling of this bridge in August 1862, two men sustained severe internal and external injuries. A board in the central arch broke while the men were working on top of the bridge, precipitating their fall through the bridge timbers to land on a pier thirty feet below. Both were hospitalized, and, though there is no record on whether they recovered, we can only imagine what became of these men given the state of medicine at the outset of the Civil War.

The fifth Wells Street Bridge reused the foundation and turntable of the previous swing bridge, and construction of a new

## Second, third, fourth, and fifth Wells Street Bridges

| Opened | Bridge type | Designed by | Constructed by | Cost | Status |
|---|---|---|---|---|---|
| July 1847 | Pontoon swing, wood, hand operated | Unknown | Unknown | $3,000 | Destroyed by flood on Mar. 12, 1849 |
| Sept. 1849 | Pontoon turntable swing, wood, hand operated | Unknown | Unknown | $3,200 | Removed in Feb. 1856 |
| 1856 | Swing, wood, hand operated | Derastus Harper, city superintendent of public works | Derastus Harper, city superintendent of public works | $20,000 | Removed in Aug. 1862 |
| 1862 | Swing, pivot, wood, hand operated | Fox & Howard | Fox & Howard | $8,000 | Destroyed by fire on Oct. 8–10, 1871 |

superstructure was completed in 1862. As heavily used as its predecessors, it was the scene of a huge traffic jam the year after it opened as five hundred people and a half mile of wagons were backed up at the bridge for forty-five minutes. It was held open for the passing of the barge *Sunny Side.* A strike by the tug captains had forced the crew to manually pull the barge through the opening using hauling lines. With the captain not aboard, the mate was in charge, and the barge was further delayed as two of the crew "managed to get up a fight."[2] Meanwhile, the best alternative, the Clark Street Bridge, was broken and had been swung open for repair earlier in the day. The only means of crossing north and south over the river was the Rush Street Bridge.

Another common problem with swing bridges occurred at the Wells Street Bridge on October 3, 1869. Just before midnight, the bridge was swung to allow a tug to pass when "a carriage approached from the south, and the horses, at quite a slow pace, proceeded toward the gap. It is presumed that the driver was either extremely drowsy or quite asleep, since he gave no sign that he was aware of the situation; and the horses without a guide and with an unusual lack of 'horse sense,' passed beyond the approach and fell headlong into the stream, the carriage and driver following in their wake."[3] The driver did not survive. It was noted that the approach to the bridge was not well lit. Preventable accidents like these would not ultimately be solved

Looking east at the sixth Wells Street Bridge with Clark Street Bridge in the background in the late 1870s.

until the advent of bascule bridges. The bascule design, which opens vertically, raises the bridge deck to present a clear barrier to street traffic from falling into the river. Various barriers for swing bridges were proposed over the next forty years, but no effective solution was ever implemented. This bridge was burned to the waterline in the Great Chicago Fire.

The sixth Wells Street Bridge required a completely new substructure and cost eight times as much as its predecessor.

## Sixth Wells Street Bridge

| Opened | Bridge type | Designed by | Constructed by | Cost | Status |
|---|---|---|---|---|---|
| Aug. 1872 | Swing, iron, hand operated | Fox & Howard | Fox & Howard | $49,002 | Moved to Dearborn Street on Mar. 26, 1888 |

View northwest of the seventh Wells Street Bridge and clock tower of the Chicago and North Western Railway's Wells Street Station prior to 1911. Courtesy of the Chicago Maritime Museum.

After sixteen years in service, the all-iron, 190-foot span was removed and reused at Dearborn Street, where it served for another eighteen years. To the utter amazement of the watching crowds, this ninety-ton structure was jacked up, floated off of its substructure, maneuvered through the open Clark Street Bridge, and positioned at Dearborn Street in the space of just two days. The fascinating newspaper report of this move is reproduced in the sidebar earlier in this section. A new steel swing bridge replaced it.

The seventh Wells Street Bridge superstructure was provided to the City by the North Chicago Street Railroad Company under their contract regarding use of the La Salle Street Tunnel. This steam-powered, 220-foot steel bridge cost the streetcar company $59,000, plus $4,690 for the operating machinery. The City paid for the substructure. In 1890 the Department of Public Works made extensive repairs to the center pier and north approach, including driving approximately fifty new 45-foot pilings. A new set of eighty cast-iron wheels was also installed on the turntable of this bridge.

On July 5, 1896, the bridge was closed so that the firm of Shailer and Schniglau could add a top deck to carry the trains of the Northwestern Elevated Railroad Company. The bridge's turntable was beefed up by adding a new set of steel rollers, a plate-girder drum, a center step, and a distributing girder. Shailer and Schniglau also removed the seven center panels from all three bridge trusses, rebuilding them with heavier steel beams that could handle the additional weight. Other improvements

| Seventh Wells Street Bridge | | | | | |
|---|---|---|---|---|---|
| Opened | Bridge type | Designed by | Constructed by | Cost | Status |
| Aug. 24, 1888 | Swing, steel, steam, then electric powered | Keystone Bridge Co. | Keystone Bridge Co. (superstructure), FitzSimons and Connell Co. (substructure) | $112,692 | Closed on Dec. 3, 1921, and then removed |

included partially repaving the lower roadway, converting the bridge to electric power, and installing a new bridge house. The work was completed, and the bridge was reopened on September 20 of that year. In 1902 the bridge house was moved from the west side to the east side of the street and repainted, and replacement of the protection pier begun in 1901 was completed. In 1907 a new danger signal system was installed that added side lights to the trusses. A year later, the old bridge house was removed, and a new one was built over the sidewalk in the middle of the bridge.

At the beginning of the twentieth century, replacing the double-deck swing bridges at both Wells Street and Lake Street presented a significant technical challenge. Still, the Department of Bridges was more than equal to the task and in 1916 installed the first double-leaf, double-deck bascule at Lake Street Bridge. The Wells Street Bridge soon followed, and, on December 3, 1921, the old double-deck swing bridge was removed and the new bascule bridge that had been erected around it was lowered into place.

# FRANKLIN-ORLEANS STREET BRIDGE

LOCATION: 230 West, 307 North; Franklin Street, which runs from the river to the south, and Orleans Street, running north of the river, were joined by this bridge. It crosses the Main Channel one and a half miles west of the river mouth at Lake Michigan.

Completed in 1920, the Franklin-Orleans Bridge is virtually identical to the Monroe Street Bridge, built in 1922, and the third Belmont Avenue Bridge, which opened in 1913. It is a second-generation Chicago-type double-leaf bascule measuring 251 feet from trunnion to trunnion and 62 feet in width, providing a 220-foot waterway when open. The twin pony-truss steel superstructure supporting the roadway is of rivet gusset-plate construction, the joining of three or more steel members by affixing steel plates to the outside of the joint using steel rivets. The abutments and bridge approaches are reinforced concrete with a rusticated concrete veneer and two granite-faced, octagonal bridge houses. The bridge houses sit on opposite corners of the bridge, and their hipped roofs of simulated tile give the tiny control rooms a solid, substantial appearance.

The Franklin-Orleans Street Bridge was constructed as part of a $1.9 million Franklin-Orleans street-improvement proj-ect. The bridge was specifically designed for heavy truck traffic and meant to relieve the Lake and Wells Street bridges. It was also expected to spur economic development and serve the warehousing district on the Northwest Side. Bridging these two streets presented several problems, however: Franklin and Orleans streets did not align at the river, proposed street improvements meant several buildings would need to be taken down, the grade had to be raised to cross over the tracks of the Chicago & North Western Railway, and the new connection would impact several streets in the area, including Lake, Kingsbury, Wells, and South Water streets. In 1916, a decade after the plan was first proposed, appropriations were finally approved, and construction began. Although World War I and the postwar construction of the Michigan Avenue Bridge delayed progress, the resulting bridge benefited greatly from the significant advancements made in bascule-bridge design during the delay.

The theme of the bridge's opening parade on October 23, 1920, was "Ship by Truck," coinciding with a national campaign begun by tire manufacturer Harvey S. Firestone. This was to be more than just a practical bridge, however, as the design significantly reflected the aesthetics of the City Beautiful movement. The grand style of the Franklin-Orleans Street and Michigan Avenue bridges, both completed in 1920, made them the first

| Franklin-Orleans Street Bridge | | | | | |
|---|---|---|---|---|---|
| Opened | Bridge type | Designed by | Constructed by | Cost | Status |
| Oct. 23, 1920 | Chicago type, double-leaf bascule | Bureau of Engineering, Chicago Department of Public Works | Keystone Bridge Co. (superstructure), FitzSimons and Connell Co. (substructure) | $1 million | Currently in use |

Looking west at the Franklin-Orleans Bridge in 2009. © Kevin Keeley

"aesthetically pleasing" bridges on the main branch of the river.[1] These bridges mark the first efforts toward the transformation of the riverfront area into a gateway to the city. Subsequent bridges built across the main branch, though not as ostentatious, would incorporate similar aesthetic considerations in their design and appearance up through the 1930s.

Firestone may well have succeeded in his goal to put more trucks on the road, as seven years later the Franklin-Orleans Street Bridge was greatly worn and in need of repair. It was re-decked with asphalt-covered timber planking. The original bridge included four decorative terra-cotta pylons housing signal lights at each corner of the bridge. In 1936, while fixing an expansion joint between the approaching viaduct and the bridge, these pylons, due to their cracked and dangerous condition, were re-placed with four arc light poles sporting new stoplight boxes. The bridge received a second new floor when a concrete-filled steel grid deck was installed in 1951. The bridge was converted to one-person operation in 1958, allowing control of both leaves of the bridge from one bridge house. The bridge has also under-gone three major rehabilitations, in 1940, 1971, and 1992.

Pylons on the Franklin-Orleans Bridge's north approach as shown in 1923. Courtesy of MWRD.

# THE BRIDGES OF THE

## (FROM NORTH TO SOUTH)

**South Branch**
*bent, 1832–40*

**Randolph St.**
*Chicago bascule, 1984*

**Madison St.**
*Chicago bascule, 1922*

**Adams St.**
*Chicago bascule, 1927*

**Metropolitan West Side Elevated RR**
*Scherzer rolling lift, 1894-1956*

**Congress St.**
*Chicago bascule, 1956*

**Chicago Terminal Transfer RR**
*(various), 1885-1929*

**Baltimore & Ohio Chicago Terminal RR**
*Strauss heel trunnion bascule, 1926*

**St Charles Air Line RR**
*Strauss heel trunnion bascule, 1919*

**18th St.**
*Chicago bascule (single leaf), 1967*

**Canal St.**
*Chicago bascule, 1948*

**S. Halsted St.**
*Chicago bascule, 1934*

**Throop St.**
*(various), 1868-1978*

**Loomis St.**
*Chicago bascule, 1978*

**S. Ashland Ave.**
*Chicago bascule, 1938*

**Lake St.** Wacker
*Chicago bascule, 1916
double-deck*

**Washington St.**
*Chicago bascule, 1913*

**Monroe St.**
*Chicago bascule, 1919*

**Jackson Blvd.**
*Strauss bascule, 1916*

**Van Buren St.**
*Chicago bascule, 1956*

**Harrison St.**
*Chicago bascule, 1960*

**Polk St.**
*(various), 1855-1972*

**Taylor St.**
*(various), 1890-1929*

**Roosevelt Rd.**
*Chicago bascule, 1930*

**Pennsylvania RR**
*vertical lift, 1914*

**Cermak Rd.**
*Scherzer rolling lift, 1905*

**Dan Ryan Expwy.**
*fixed, 1965*

Map of the South Branch of the Chicago River. © Chicago CartoGraphics

The South Branch of the Chicago River stretches from the fork of the Main Channel south five miles and then splits into the South and West forks. Throughout Chicago's first hundred years, the vast majority of commercial businesses, heavy industry, and urban population were centered on this section of the river. Situated between the Main Channel and South Ashland Avenue, the South Branch hosts nineteen highway bridges and two railroad bridges. This section of the river has been spanned only by two fixed bridges: the South Branch Bridge, built in 1833, and the Dan Ryan Expressway Bridge, opened in 1965. All other South Branch bridges are required by federal oversight to be moveable. The South and West forks of South Branch have been home to a mixture of fixed and moveable bridges, as dictated by the needs at the time.

The South Branch has undergone the most significant modifications of all of the city's waterways to better serve the commercial and industrial needs of Chicago. Modifications in the early 1860s added more than fifteen industrial slips, each two to three

# SOUTH BRANCH

blocks in length, to attract industry to the area. Originally just a creek, the West Fork flowed east from the street line of Western Avenue into the South Branch. It was widened and extended in the 1850s through a series of ditches reaching west to the Des Plaines River, which were meant to improve drainage of the surrounding lands. In 1871 a fifty-foot-wide canal was initiated by William Ogden and John Wentworth known as the Ogden-Wentworth Ditch (or Canal). The canal added to development of the adjoining lands but created backflow problems that affected the Chicago River and I&M Canal. Protest from the city led to a dam being added to block flooding from the Des Plaines in 1877.

The South Fork of the South Branch, now only one and a half miles long, originally extended another half mile south to about 39th Street (renamed Pershing Road) and another mile west to Western Avenue. Dredged and widened in the 1860s for shipping, a mile-long eastern channel was added in line with Pershing Road along the northern edge of the Union Stock Yards. This collateral canal was maintained into the 1930s and served

the city's slaughterhouses, giving the South Fork the moniker *Bubbly Creek.* The decomposition of by-products from decades of dumping animal waste into the South Fork caused the water to bubble, which it still does today despite the official close of the Union Stock Yards on July 30, 1971.

From 1928 to 1930, the waterway between Polk and 18th streets was straightened. This immense project replaced an easterly bend in the river with a mile-long channel in line with Clark Street. Planning and implementation involved a complicated transference and reallocation of property between the City, the railroads, and various other landowners. Initially proposed by Daniel Burnham in 1907 as part of the *Plan of Chicago*, it was first officially recognized in the Union Station Ordinance of 1914. The total cost of the project was approximately twelve million dollars, and it involved the largest and most intricate real-estate deal ever attempted. Involving the survey of land, documentation, and legal work necessary to transfer about 120 lots ranging in size from 28 to 280,000 square feet, each parcel of land had to be transferred twice, and plural ownership, riparian rights, and various other legal issues complicated the deal. Richard Wolf, the commissioner of public works, led the effort, with the significant aid of the project trustee, the Foreman-State Trust & Savings Bank. The project also involved the relocation and construction of several bridges, particularly the Roosevelt Road Bridge. The legal issues were settled by September 1928, and the main contractor for the project, Great Lakes Dredge & Dock Company, dug the new channel and filled in the old river channel in just twenty-two months.

To this day, the South Branch of the Chicago River is still lined with rail lines and commercial, warehousing, and manufacturing buildings (many now repurposed) in a testament to its industrial past. This branch holds an interesting variety of drawbridge designs that are often ignored because of their location in this industrial area.

Aerial view north of the old channel of the South Branch of the Chicago River in 1925. Courtesy of the Chicago Maritime Museum.

# LAKE STREET BRIDGES

LOCATION: 356 West, 200 North; Lake Street runs east and west and crosses the South Branch, just below the forks, 1.6 miles from the river mouth.

HISTORICAL HIGHLIGHT: Today's Lake Street Bridge was the first double-deck, double-leaf bascule bridge in the world. It provides a model for the subsequent double-deck Michigan Avenue and Wells Street bridges. Chicago's very first swing bridge was built at Lake Street in 1852.

Today's Lake Street Bridge, the fifth at this location, was precipitated by a federal order for removal of the swing bridge that preceded it. Like Wells Street, the old bridge had been modified with the addition of an upper deck carrying elevated trains in and out of the Loop. City engineers prepared three designs—a vertical-lift and two different bascule bridges—to potentially replace the existing swing bridge. Reviewed by chief engineer John Ericson, the Oak Park Elevated Railroad's chief engineer, and a third independent bridge engineer, the vertical lift was selected as the least complex of the three proposed designs.

This news was quickly met with outrage from the Chicago Plan Commission and related civic groups, however, leading to the ultimate rejection of the unsightly vertical-lift design. City engineers rose to the challenge and designed a double-leaf, double-deck bascule bridge, the first of its kind in the world. The total length of this bridge is 355 feet, with 245 feet between the trunnions. It is 70 feet wide, and the entire structure rests on concrete subpiers reaching 110 feet down into solid rock. When open, the bridge provides an unobstructed river channel of more than 200 feet. Erection of the bridge took nearly two years.

This new Lake Street Bridge solved several major technical and design challenges so that it could carry elevated-train traffic on its upper deck and street traffic on the lower deck. The Lake Street location held additional complications, however. City engineers were required to accommodate both a set of railroad tracks just below street level along the west riverbank and a planned double-track subway tunnel that would run through the bridge's foundation. The bridge, twice as heavy as any previous city bridge, was constructed in the upright position so that the old swing bridge could continue to be used at this busy crossing.

| | | **Current (fifth) Lake Street Bridge** | | | |
|---|---|---|---|---|---|
| Opened | Bridge type | Designed by | Constructed by | Cost | Status |
| Nov. 6, 1916 | Chicago type, double deck, double-leaf bascule | City engineers John Ericson, Thomas Pihlfeldt, Alexander von Babo, and Hugh Young and architect Edward Bennett | Ketler-Elliott Co. (superstructure), FitzSimons and Connell Co. (substructure) | $610,000 | Currently in use |

Looking south at the open current Lake Street Bridge in the spring of 2009. © Kevin Keeley

Traffic on the busy street and the 494 train crossings per day were maintained, and the estimated 200 swings per day allowed ship traffic to pass during construction.

Once construction neared completion on February 27, 1916, elevated-train traffic was halted, and crews worked night and day to minimize the interruption to the commuter trains. They removed the old swing bridge, lowered the leaves of the new bridge, completed final portions of the upper deck, and installed the elevated-train tracks. By the afternoon of March 4, bridge trains were back on their regular schedule, crossing the new flag-and-bunting-decorated bridge. Thomas Pihlfeldt, with the cooperation of engineers from the Chicago & Oak Park Elevated Railroad Company, orchestrated the transfer and resumption of "L" traffic onto the new bridge within a week; the lower deck remained closed to street traffic for several months, however, while it was being completed. The rail company, by contributing

Construction of the current Lake Street Bridge in 1915.

one-third of the construction costs, retained use of the Lake Street Bridge's upper deck.

As an interesting side note, the Chicago Transit Authority, which now runs "L" trains over Lake Street, controls the electricity to operate this bridge. CTA personnel must first throw a track switch so any runaway trains are then routed to a dead-end bumper in front of the bridge. The track switch also activates a limit switch to turn on the power to bridge operation. These safety measures ensure cooperation between the CTA and the Chicago Department of Transportation, to prevent train traffic from attempting to cross during a bridge lift.

In 1926 on the east approach, and then in 1931 on the west approach, the original boom safety barriers that automatically lowered into place to stop street traffic were removed and replaced with a yielding barrier. This was installed to avoid the risk of taking the tops off of vehicles that hit the boom before it was completely lowered, as had been experienced by two automobiles at Clark Street. Reopened to traffic on September 22, 1930, the lower deck of this bridge received sixteen different kinds of pavement as a test of relative durability. The trial surfaces included asphalt planks, rubber tile, creosoted blocks, oak planks on wood subfloors, "Tymber" brand slabs, "Unicate" brand mastic, battle-deck steel, and precast concrete slabs welded into place. No conclusive results of this test were ever publicly reported, and in 1950, the bridge-floor system and its steel supporting structure were entirely rebuilt. A concrete slab roadway was installed on the bridge approaches, and the ends of the bridge deck were replaced with an I-Beam-Lok, concrete-filled surface; the other half of each bridge leaf received one-and-a-half-inch mineral-surfaced asphalt planking over three-by-six-inch decking resting on six-by-twelve-inch

## First Lake Street Bridge

| Opened | Bridge type | Designed by | Constructed by | Cost | Status |
|---|---|---|---|---|---|
| 1852 | Pivot, wood, hand operated | Streets superintendent, Derastus Harper | Streets superintendent, Derastus Harper | Unknown | Removed in 1859 |

subplanks. Major structural repairs were made to this bridge in 1959, and in 1969 it was repaired and entirely redecked with an open steel-grid roadway.

The very first bridge at Lake Street was also the very first center-pier pivot bridge constructed in Chicago. Derastus Harper designed this bridge, and he voluntarily rescinded half of his salary so he could act as its contractor. The bridge was initially left open at night, but the Common Council soon approved a proposal by the Harbor and Bridges Committee that the "bridge be attended to all night, the same as the other bridges for the convenience as well as the safety of the public."[1] Subsequently, under City approval, Harper added a few "snubbing" piles to prevent ships from being driven against the bridge, particularly during a north wind.[2] These initial pilings were the genesis of the extensive protection center piers, first installed at the third Clark Street Bridge, that became standard on all swing bridges. The protection piers extended just beyond the outline of the open bridge in the center of the river channel to fend off passing ships.

In early 1857, this pivot bridge received seventeen hundred dollars in improvements, though this did not include a proposed increase to the clearance to allow for the passage of tugs and canal boats. The repairs were paid for with both private and City funds. Several months later, on December 3, the steamer *Foster* hit and broke one of the bridge's supporting cables. The bridge was then swung open for more than a week while repairs were made. This first pivot bridge was removed in 1859 after seven years of service.

Looking northeast at the second Lake Street Bridge in 1866.

| Second Lake Street Bridge | | | | | |
|---|---|---|---|---|---|
| Opened | Bridge type | Designed by | Constructed by | Cost | Status |
| July 2, 1859 | Swing, wood, hand operated | Newton Chapin | Chapin & Co. | $14,560 | Removed in 1867 |

Looking northeast at the third Lake Street Bridge in 1879.

In January 1859, the Common Council advertised for bids on a second Lake Street Bridge. Awarded to Chapin & Company, a 185-foot bridge resting on a pile center pier and stone abutments was built. A substantial structure for its day, it opened on July 2, 1859, and on July 7, in an article titled "Another Bridge Accident," the *Chicago Tribune* reported: "Last evening, the bark *W. F. Allen* ran into the Lake Street bridge, while being towed up the river, and her jibboom [was] carried away. Accidents of this nature are becoming very frequent, and in five cases out of six, are entirely owing to the stupidity and want of common sense on the part of the bridge-tenders. Will our city fathers appoint men of intelligence to fill these positions, instead of giving them to individuals, whose only merit consists in being able to stuff a ballot-box?"[3] One of the first regular City jobs, bridge tending was a patronage job by the 1860s. Mistakes in operating Chicago's massive bridges were often costly. This Lake Street Bridge survived for eight years, until 1867, when it too was replaced by a new bridge.

The third Lake Street Bridge used the repaired substructure from the second, and a new bridge was built in 1867. It was a Howe truss, 33 feet wide, and the same length as its predecessor.

Sadly, in those days, Chicago's river bridges were often the scene of suicides, such as the one reported at this bridge on July 28, 1871, in an article titled "A Leap into the River":

## Third Lake Street Bridge

| Opened | Bridge type | Designed by | Constructed by | Cost | Status |
|--------|-------------|-------------|----------------|------|--------|
| 1867 | Swing, wood and iron, hand operated | Fox & Howard | Fox & Howard | $11,450 | Removed in 1885 |

*About 10 minutes after 9 o'clock last evening a young woman named Emma Cashin jumped into the river from the Lake street bridge and was drowned. She was intoxicated, and about 9 o'clock was in the saloon No. 250 Lake street. She asked for a glass of beer, but the proprietor refused to give it to her. She left seemingly good-natured, and the next seen of her was by two women who were crossing Lake street bridge. She was in the center of the bridge, throwing off her bonnet and shawl. The two women screamed and the bridge-tenders were on hand just as she jumped into the water. A boat was procured, and she was caught as she came up the third time, but life was already extinct. The body was taken to the Armory Dead House, and the Coroner notified to hold an inquest. The suicide was about 22 years of age, and had been known to the police of this city for about three years. She lived in Connelly's patch, corner of Franklin and Quincy streets, when she was not in the Armory [Court and South Side Police Station] or the Bridewell [the prison at West 26th Street and California Avenue]. Two days ago she was released from the latter institution, and had been drunk continually since. She was very good looking at one time, but whiskey gave her a bloated and repulsive countenance. Her mother resides somewhere on the North Side.[4]*

Thanks to the heroics of bridge tender Martin Casey, this bridge escaped destruction by the Great Fire of 1871. He held the bridge to get people across, despite repeated orders to swing the bridge. Some fifty men broke into a nearby hardware store and secured shovels, pry bars, and buckets to drench the bridge with water, dig trenches, and tear up burning approach planks, fighting the flames to maintain the crossing. They did not cease until every last soul was across. These heroic men at Lake Street and other bridges combined with the tunnels under the river at La Salle and Washington streets saved a great many people from the inferno.

In 1881 the West Division Railway Company paid to modify the bridge by moving the trusses apart and widening the roadway to accommodate streetcars. Four years later, the third Lake Street Bridge was removed. Dismantling the bridge was straightforward, but removing the center pier proved slow and tedious. Formerly supporting the bridge and turntable were about 250 pilings 45 feet in length driven deep into the riverbed. Two pile-driving barges were used to remove these pilings. One operated a circular saw, driven by and attached to a long shaft, to cut the pilings 17 feet below the water surface. Then, a second pile driver pulled each piling out of the river. The shortened pilings left in the river were then used as a foundation for a new caisson supporting the next bridge.

In 1885 the West Division Railway Company contributed half of the cost for the fourth Lake Street Bridge to ensure that it would carry its streetcars across the South Branch. This

Looking southwest at the fourth Lake Street Bridge during construction of a new bascule bridge in 1915.

119

wrought-iron, seven-hundred-ton Pratt-truss bridge was similar in design to the fourth Rush Street Bridge, which was also built by the Detroit Bridge Company, a year earlier. This Lake Street Bridge was 220 feet long and 59 feet wide, with two roadways that were each 21 feet wide. Each roadway carried streetcar tracks, with 8½-foot-wide sidewalks off to each side. The bridge turned on a 48-foot diameter drum containing eighty wheels 18 inches in diameter and 8 inches thick revolving around a steel track. The bridge machinery consisted of two locomotive-type boilers that provided the steam for the twin engines that swung

## Fourth Lake Street Bridge

| Opened | Bridge type | Designed by | Constructed by | Cost | Status |
|---|---|---|---|---|---|
| June 1886 | Swing, iron, steam, then electric powered | Detroit Bridge Co. | Detroit Bridge Co. (superstructure), FitzSimons and Connell Co. (substructure) | $127,259 | Removed in Feb.– Mar. 1916 |

the bridge. Once the bridge was closed, two additional engines wedged the ends of the bridge into place, and a final engine pumped water from the river for engine cooling and from the water mains for the boilers. Produced by the Vulcan Iron-Works of Chicago, the engines had all-steel shafting and bronze gearing. The same firm provided similar machinery for the Rush Street, the Chicago & Evanston Railroad, and the Chicago & Great Western Railroad bridges in Chicago. An additional steam engine drove the electric dynamo, powering seven electric lights of the Van Dopoele Electric Manufacturing Company to "handsomely" light the bridge.[5] Van Dopoele was the preferred exterior lighting manufacturer at the time.

This bridge was converted to electric power in 1893 after "extensive and rather remarkable alternations."[6] The competing Lake Street Elevated Railroad paid to reinforce the structure, which at its center used a Warren-truss configuration, and added an upper deck to carry elevated trains. The project gave the company an all-important connection into and out of Chicago's Loop. Charles L. Yerkes, the financier and Chicago traction baron, would acquire both companies in 1894, ending the rivalry. In 1896 new subplanking was installed, and the bridge deck was repaved. In 1902 the Lake Street Elevated Railroad installed a new G.E. 800 motor to replace one that had burned out in September. The City next purchased two new G.E. 58 motors, which were installed in early 1909. That same year, the U.S. Army Corps of Engineers deemed the bridge "an unreasonable obstruction to the free navigation of the Chicago river on account of the center pier, narrow draw openings and faulty location."[7] The secretary of war ordered its removal and replacement with a vertical-lift or bascule bridge with a minimum 16½-foot clearance. In 1916 this double-deck swing bridge was removed and replaced by the first-ever double-deck, Chicago-type bascule bridge.

# SOUTH BRANCH BRIDGE

(No Current Bridge)

LOCATION: There has not been a bridge there for more than 160 years, but a South Branch Bridge crossed the river just north of Randolph Street prior to the organization of city streets.

HISTORICAL HIGHLIGHT: This was the second bridge ever built in Chicago, and the first and only bridge crossing the South Branch between 1832 and 1838 (hence its name).

With Chicago not yet a village, the first South Branch Bridge was built by two brothers, Charles and Anson Taylor. The bridge was sturdy enough to carry wagons, and the cost of the bridge was paid through citizen subscriptions. Most notably, the Pottawatomie Tribe contributed $200 of the $486.20 collected for its construction. Residents claimed the Pottawatomie got the best use of it as a diving and swimming platform for frolicking in the river. The bridge was built in the winter of 1832–33, from timber cut from the adjoining woods. Interestingly, like Chicago's first bridge, the South Branch Bridge was also associated with a tavern; in June 1832, Charles Taylor and his wife, Mary, rented the Wolf Point Tavern and ran it over the course of the next year. This new bridge would certainly have benefited Taylor's new business.

Most of what we know about this bridge is pieced together from early petitions to the Common Council. In January 1836, citizens petitioned for removal of the heavily used bridge, which was "much decayed and in a ruinous state and condition."[1] The consensus of the citizenry was that a new bridge in line with Randolph Street was needed. The Common Council took out an advertisement in the local newspaper for bids to build a moveable bridge sixteen feet wide that was "sufficient for the passing of two carriages," a draw of forty feet, and minimum clearance above the high-water mark of nine feet to allow for the "passage of smaller craft" under the bridge.[2] On February 17, 1836, J. F. Brown proposed just such a bridge, and on April 13, 1836, the *Chicago Democrat* reported that the town trustees had "decided upon building a bridge across the South Branch on Randolph street and across the north Branch on Kinzie street."[3] Neither proposal was actually undertaken.

The Bridges Committee reported on May 11, 1837, that the South Branch Bridge was unreliable and unsafe for the passage of teams or loaded wagons. Over time the bents of the bridge had sunk into the river, and during high water scows could no longer pass underneath. Adding timber to the tops of the bents to raise the bridge and make it "passable for the season" was recommended.[4] The bridge was repaired, but this did not solve

| South Branch Bridge | | | | | |
|---|---|---|---|---|---|
| Opened | Bridge type | Designed by | Constructed by | Cost | Status |
| Winter of 1832–33 | Fixed, bent bridge | Charles & Anson Taylor | Charles & Anson Taylor, helped by Fort Dearborn troops | $486 | Removed in Mar. 1838 |

the problem, as ultimately the fixed bridge still blocked the passage of ships down the South Branch.

In January 1838, Nelson R. Norton, builder of the Dearborn Street drawbridge, requested removal of the South Branch Bridge. This was done to allow passage of a work scow and pile-driving barge to construct a wharf just south of the bridge in front of the furnace works of Stow & Company on the west bank of the South Branch near Polk Street. On March 1, 1838, a later petition confirmed the bridge's removal and requested the relocation of the approaches for use at Randolph Street by a ferry. The approaches remained where they were, however, and for the next year the ferry operated at the old crossing instead. At least nine men petitioned the council to attend the South Branch ferry day and night, and proposed salaries ranging from $1.75 to $2.87 per day. In 1838 the street commissioner reported that the ferry over the South Branch was inoperable, as the rope was so worn it had completely given way. With the assistance and pledge of alderman and treasurer George W. Dole, a chain was procured and a windlass was employed to stretch it across the river and return the ferry to service. The April 11, 1839, *Chicago Daily American* reported a motion by Fifth Ward alderman John Murphy Jr. for removal of the last timber piles of the old South Branch Bridge. The motion passed, bringing an end to this crossing.

# RANDOLPH STREET BRIDGES

LOCATION: 358 West, 150 North; Randolph Street runs east and west and crosses the South Branch 1.7 miles from the river mouth at Lake Michigan.

HISTORICAL HIGHLIGHT: The current Randolph Street Bridge is the newest drawbridge built in Chicago and, along with Clark and Wells streets, has had the most bridges, with eight at this location.

The eighth Randolph Street Bridge, a modern Chicago-type bascule, took three years to complete and used the steel box-girder construction introduced in 1978 at the Loomis Street Bridge. The welded-steel construction and single bridge house give it a sleek, clean look. With a twenty-one-foot clearance, this bridge provides an additional five feet over the river and less "low steel" than the old Scherzer rolling-lift span it replaced. As a result, it opens fewer than 100 times per year, compared to 1,450 openings for the old bridge. State and federal funds paid for this bridge, which was designed by the City's Public Works Department and Hazelet & Erdal. Ironically, this Chicago engineering and consulting firm, established in 1936, is the direct successor to the Scherzer Rolling Lift Company that designed the previous Randolph Street Bridge.

The first Randolph Street Bridge was Chicago's first pontoon-float swing bridge. The rendering shown is based on plans for this bridge submitted to the Common Council in January 1839 by John Van Osdel. This bridge, constructed of oak and pine, was completed on May 6, 1839, and used a seventy-two-foot floating draw that rotated on pivots at one of two corners of the pier. This arrangement allowed the bridge to be opened up- or downstream, as needed. It was suggested that the Common Council appoint a bridge tender and install a sign warning persons not to drive over the bridge faster than at a walking pace. A bridge tender was appointed at a pay rate of $0.75 per day. In 1846 the bridge received new floats; in April 1847, it was removed to make way for a new bridge.

The second Randolph Street Bridge was the same design as its predecessor, with the exception of stone approaches added on the riverbanks. Half the cost of construction was paid through local subscriptions, and the other half came from City coffers. It was one of only five bridges in Chicago in 1847. All five bridges came to an untimely end in the flood of March 1849, as recounted in the sidebar in the introduction.

A replacement bridge was built quickly and opened five months later. The third Randolph Street Bridge was lower in elevation than its predecessor, and so needed to be opened to allow passage of almost every size craft. Approximately one-third of the expense of this bridge was paid for by subscriptions, and the City borrowed the funds to pay for the rest. It was a refur-

| Current (eighth) Randolph Street Bridge | | | | | |
| --- | --- | --- | --- | --- | --- |
| Opened | Bridge type | Designed by | Constructed by | Cost | Status |
| Dec. 18, 1984 | Chicago type, double-leaf bascule | City Bureau of Engineering and Hazelet & Erdal of Chicago | Kenny Construction Co. | $18 million | Currently in use |

Looking north at the current Randolph Street Bridge in 2009 with the Wells Street Bridge in the background. © Kevin Keeley

| First, second, and third Randolph Street Bridges | | | | | |
|---|---|---|---|---|---|
| Opened | Bridge type | Designed by | Constructed by | Cost | Status |
| May 1839 | Pontoon swing, wood, hand operated | John Van Osdel | Charles Grog, city street commissioner | $1,400 | Removed in 1847 |
| 1847 | Pontoon swing, wood, hand operated | Unknown | Unknown | $5,000 | Destroyed by flood on Mar. 12, 1849 |
| Aug. 1849 | Pontoon float swing, wood, hand operated | Bridge Committee, City of Chicago | Bridge Committee, City of Chicago | $924 | Moved to North Avenue in 1856 |

bished version of the first bridge, reconstructed with salvaged timbers from the original.

In January 1850, this bridge was improved under the oversight of Derastus Harper, who converted it into a turntable pontoon swing bridge at a cost of fourteen hundred dollars. Two larger floats were added under the draw, and the bridge was now operated using a turntable that rested on the stone foundation of the east approach. A semifloating apron was added to the west approach that connected with the floating draw and allowed for the changing water levels. The bridge had two 8-foot tracks designed for teams and raised sidewalks with railings on each side. It was furnished with operating wheels and chains, and like its predecessors could be opened up- or downriver, as the occasion required. At that time, it was considered "the best and most durable bridge there is or ever had been built in the city."[1] The bridge was removed from this important crossing in 1856 to make way for a new bridge. The old structure was reused to establish a new river crossing at North Avenue.

The fourth Randolph Street Bridge was a pivot bridge, only the fourth such design built in Chicago. This bridge was 162

| Fourth Randolph Street Bridge | | | | | |
|---|---|---|---|---|---|
| Opened | Bridge type | Designed by | Constructed by | Cost | Status |
| July 1856 | Swing, wood, hand operated | William Boomer | Stone, Boomer & Bouton | $20,811 | Broken beyond repair, May 11, 1864 |

feet in length and had three wood Howe trusses. The bridge offered two 18-foot-wide roadways, including 7-foot-wide raised sidewalks on each side. The arched trusses were 14 feet high at the middle and curved down at the ends to a height of 4½ feet. The top and bottom chords of each truss were 5" × 12" wood beams. Between these were wood crossbeams and vertical iron compression rods that ran through the chords. Lateral overhead cross-bracing at the center of the three trusses provided additional strength. The two roadways had 3-inch-thick planking, and there was 1½-inch-thick planking on the sidewalks.

The approaches and center pier were built on pilings driven into the riverbed. The center pier consisted of a clump of pilings at the center and a 2-foot-wide circle of pilings wrapped together by ¾-inch chain. The inside was filled with stone up to the low-water mark and encircled by eight more pilings in a 27-foot circle. The outer circle of pilings was capped with 12" × 12" timbers and four additional 6" × 10" oak timbers that tied the center pier's pieces together. This was covered by 5-inch planking, which also held the bridge turntable. The three-truss bridge superstructure was manually operated with the aid of turntable gears. The superstructure received three coats of white lead and linseed oil paint, and the turntable was given two coats of Ohio paint.

The early wood pivot bridges typically lasted six to eight years, about a third longer than the early pontoon swing bridges. The Randolph Street Bridge lasted eight years until one day in 1864 when the roadbed parted in the middle. An uneven rent, ½-inch wide, appeared, and the "manly" bridge tenders were praised for continuing to perform the "rather dangerous task" of swinging the bridge that afternoon. At about four thirty in the afternoon, the streetcar company halted its cars, and the bridge

Looking north at the fifth Randolph Street Bridge, with Lake Street Bridge in the background, in 1870.

was swung open one last time to allow the free passage of ships. This allowed the bridge to be pulled down before it fell into the river. As the paper reported, this bridge was treated like an overly fat hog and "killed to save his life."[2]

In the meantime, a fifth bridge at Randolph Street was being constructed off-site over a five-week period. The new bridge was installed using the old foundation and was opened to travel in about two weeks' time. This bridge was 153 feet long and 32½ feet wide, and the double-truss structure lasted two years longer than the previous bridge. It would prove a godsend, carrying thousands of citizens across the river to escape the flames of the Great Fire of 1871. By 1874, however, this bridge had exceeded its expected life span and was torn down to allow for construction of a new bridge.

The sixth Randolph Street Bridge was an all-iron bridge that weighed 134 tons and was 157 feet long by 34 feet wide. It too reused the existing foundation after a few repairs were made to the old work. This was the very first steam-powered bridge in Chicago. The design was similar to the second State Street Bridge built two years earlier. In 1897 the Randolph Street Bridge was converted to electric power by the City Bridge Department. The iron bridge superstructure served for twenty-nine years before it was removed for a new bascule bridge. This bridge lasted much longer than most, as the average life span of an iron swing bridge in Chicago was twenty-one years.

The Chicago Sanitary District built the seventh Randolph Street Bridge, which measured 169 feet in length and 72 feet in width. Following decades of limited bridge funds, the City received this Scherzer rolling-lift bridge from the Sanitary District.

This Randolph Street Bridge was subject to significant repairs, including frequent replacement of rivets on the soleplates and track girders, bridge-deck reinforcement in 1909, and repaving in 1911, 1916, 1920, and 1926. The segmental steel girders were replaced in 1926, and a new bridge house and electric gates

| Fifth, sixth, and seventh Randolph Street Bridges | | | | | |
| --- | --- | --- | --- | --- | --- |
| Opened | Bridge type | Designed by | Constructed by | Cost | Status |
| July 1864 | Swing, wood, hand operated | L. B. Boomer | L. B. Boomer | $5,000 | Removed in 1874 |
| 1874 | Swing, iron, steam powered | Keystone Bridge Co. | Keystone Bridge Co. | $10,850 | Removed in 1903 |
| Apr. 15, 1903 | Scherzer rolling lift, steel, electric powered | Scherzer Rolling Lift Bridge Co. | American Bridge Co. (superstructure), Jackson & Corbett (substructure) | $254,397 | Removed in July 1981 |

Looking west at the seventh Randolph Street Bridge under construction in 1903. Courtesy of MWRD.

were installed in 1931. Extensive structural rehabilitation was performed in 1942 and 1959, and both rehabilitations included a new bridge deck, the latter composed of a concrete-filled steel-grid roadway. In May 1955, the bridge was converted to one-person operation. After nearly eighty years, in July 1981, "the old lift bridge built for horse and buggy traffic" was demolished, and the Randolph Street crossing was closed for more than three years while the current bridge was being constructed.[3]

# WASHINGTON STREET TUNNEL AND BRIDGES

LOCATION: 364 West, 50 North; Washington Street runs east and west, crossing the South Branch 1.8 miles from the river mouth at Lake Michigan.

HISTORICAL HIGHLIGHT: This location was one of three that offered both a bridge and a tunnel, connecting citizenry over and under the river.

The current Washington Street Bridge is a second-generation Chicago-type bascule bridge that was completed in 1913. It is 197 feet between the trunnions and 57 feet wide; when open, it offers a 170½-foot clear channel. The span has two riveted steel pony trusses and uses the internal rack-and-pinion drive invented by Alexander von Babo. A 1911 bridge bond issuance by the City financed the building of this and four other double-leaf bascules and one single-leaf bascule bridge. Although they have been significantly altered over time, the bridge houses at Washington Street continue to feature wood side panels and hipped tin roofs. A hipped roof describes a roof where all four sides slope downward to the walls, forming a pyramid shape. This kind of roof requires a more complex system of trusses but gives the bridge house a formal, solid, and compact appearance.

Architecturally, this type of roof is highly durable and projects comfort and solidity.

In 1917 maintenance of this bridge included removal of the old Shuman pavement and subplanks and installation of a solid floor of 6" × 12" creosoted yellow pine overlaid by 3½-inch creosoted tamarack block pavement. The east approach Shuman Slip-Not pavement was removed and replaced with a 5-inch sandstone block laid in a dry sand and cement mixture. Based on this trial run, the railroads and City agreed to use sandstone on all approach grades over 2 percent. From 1943 to 1944, the bridge was redecked and received structural steel repairs and an overhaul of the electrical and mechanical equipment. It was converted to one-person operation in 1962 and redecked again in 1965. The City continues regular inspection of the bridge and makes repairs as needed, but more recent bridge maintenance records are very difficult to obtain since the discontinuance of the Chicago Department of Public Works Annual Reports in 1970.

After three years of debate, the City completed a tunnel connecting street traffic under the river at Washington Street on January 1, 1869, at a cost of $512,708. The original design, proposed by J. J. Gindele, a civil engineer and president of the Board of Public Works, was reviewed and modified by city engineer Ellis Chesbrough to arrive at the final design.

| Current (second) Washington Street Bridge | | | | | |
|---|---|---|---|---|---|
| Opened | Bridge type | Designed by | Constructed by | Cost | Status |
| May 26, 1913 | Chicago type double-leaf bascule | City Bureau of Engineering and architect Edward Bennett | Strobel Steel Construction Co. (superstructure), FitzSimons and Connell Co. (substructure) | $238,288 | Currently in use |

Looking north at the current Washington Street Bridge with the Randolph Street Bridge and Wells Street Bridge in the background in 2008. © Laura Banick

The resulting tunnel followed the centerline of Washington Street between Franklin and Clinton streets. Three arches built under the river encased the two roadways, each 11 feet wide by 15 feet high, and a pedestrian walkway 5 feet above the roadway grade. The tunnel was constructed in two phases using a cofferdam that was half the width of the river. Once the cofferdam was in place and the water pumped out, a trench in the riverbed was dug and a 7-foot concrete foundation and a subtunnel for drainage were laid. Next, 22-inch-thick masonry tunnel walls and arches were constructed and coated on the outside with two layers of common lime mortar sandwiching a waterproofing layer of asphalt mastic. Another 10 inches of limestone flagging was added to the top to protect the tunnel from dragging ship anchors.

The structure was designed to safely withstand almost any eventuality, even the sinking of a vessel loaded with iron. The

Looking west at the Washington Street Tunnel's east approach before its use by streetcars in 1889. Courtesy of CDOT.

tunnel extended for 222 feet between the covered approaches to either side. The tunnel approaches were protected from flooding by 8-foot-high solid-stone walls at the dock lines and 4- to 8-foot-high retaining walls around each side of the approach openings. The roadway and approaches were paved with wood blocks. The east roadway had a 6.25 percent grade from Franklin Street to the double-driveway slopes, which then had a 2.33 percent grade to the center of the river. The west approach had a 5.4 percent grade up to Clinton Street.

Leaking plagued the tunnel after its opening, however, likely caused by problems introduced during construction. Asphalt mastic, mixed under the supervision of Army Corps of Engineers on Governor's Island in New York and sold to the City of Chicago at cost, was applied during freezing weather; this is believed to have caused an incomplete seal. This problem was compounded by the shifting weight of the riverbed and at least two ship collisions with the cofferdam, both of which created cracks in the masonry during construction of the tunnel.

| Washington Street Tunnel | | | | | |
|---|---|---|---|---|---|
| Opened | Tunnel type | Designed by | Constructed by | Cost | Status |
| Jan. 1, 1869 | Arch masonry and concrete | J. J. Gindele and Ellis Chesbrough | J. L. Lake | $512,708 | Removed in Jan. 1953 |

The dark tunnel made for a less than desirable means of crossing the river. The tunnel was plagued with leaks in the ceiling, and the resulting water and ice in the tunnel created a maintenance headache for the City. Regardless, the Washington Street Tunnel provided an important means of escape for Chicagoans during the Great Fire.

In February 1886, the City gave the Chicago Passenger Railway Company permission to lay tracks and operate in various new locations, including the Washington Street Tunnel. Though the company initially used horse-drawn streetcars, they were permitted to use cable power two years later under the condition that they also make repairs to the tunnel in accordance with City specifications.

On March 15, 1889, the West Chicago Street Railway leased the lines of the Chicago Passenger Railway Company, and by June 1890, the tunnel at Washington Street was part of their Milwaukee Line. The Milwaukee Line looped from Wells to Madison, Madison to La Salle, La Salle to Randolph, Randolph to Wells, and Wells to Washington. From Washington it ran west through the tunnel under the Chicago River to Des Plaines Avenue and then up Milwaukee Avenue to Armitage Avenue. The Wells, Randolph, La Salle, and Madison loop was also shared by the Madison Street line and would eventually be elevated, becoming part of today's elevated train system, or "L."

Because of the planned reversal of the flow of the Chicago River, larger ships required a deeper river for passage. The tunnels allowed a depth of sixteen feet, but a new minimum river depth of twenty-six feet was proposed by the Sanitary District. In June 1899, the Chicago Union Traction Company acquired the West Chicago Street Railroad Company, and several months later, the City passed an ordinance requiring the company to lower the Washington Street Tunnel within three months. When the ordinance was disputed, Congress stepped in on April 27, 1904, and passed an act declaring the tunnels an obstruction to free navigation and directing the secretary of war to give notice to lower or remove them to provide a river depth of at least twenty-one feet. On June 18, 1906, the City passed an ordinance in agreement with the receivers of the Union Traction Company to lower all three Chicago tunnels. The ordinance stipulated that the City would have the option to purchase the entire system at a fixed value in anticipation of developing a citywide subway system, while the railways would continue to operate and pay for the lowering of the tunnels. Shortly thereafter, Angus Brothers & Company proceeded with the work. The tunnel was closed on August 1, 1906, and reopened to electric streetcar traffic in 1911.

In 1930 a sand barge, the *Michigan,* owned by the Great Lakes Dredge & Dock Company, ran aground on the roof of the streetcar tunnel and became stuck. The Washington Street Bridge, which had opened for the barge at ten thirty at night, had to remain open until the *Michigan* was freed. The incident occurred just as many motorists were heading home after the opera, and crowds gathered to watch the efforts made to release the large craft. Low river levels, combined with the grounding on the tunnel, resulted in a snarl of street and river traffic for several hours, despite the efforts of two tugs to pull the *Michigan* free. Experienced river men at the scene recommended closing the locks at Lockport to raise the water level and allow the *Michigan* to float free. Meanwhile, a couple of streetcar supervisors, Charles Doherty and James McCarthy, were given the unenviable

Preparing the first Washington Street Bridge to be moved prior to reconstruction of the Washington Street Tunnel in 1906. Courtesy of the Chicago History Museum.

job of watching the roof from inside the tunnel, with instructions to halt all streetcar traffic at the first sign of a leak. After several tense hours, the ship was freed without any apparent damage to the tunnel, and both river and street traffic resumed.

This tunnel was used until January 1953, when streetcars were rerouted to Madison Street to establish a consistent pattern of one-way streets downtown. The tunnel was closed, and a year later the east approach was filled in, permanently ending its use.

The first Washington Street Bridge, built in 1891, utilized the old Madison Street Pratt-truss bridge first constructed in 1875. The substructure was constructed by the West Chicago Street Railway Company at a cost of $46,099 and installed after repairs were made to the Washington Street Tunnel. Initially steam driven, this bridge used electricity for the first time on April 15, 1898. Two years later, the bridge was entirely repaved. In January 1907, this bridge was moved by scow to a city yard for storage so that the tunnel below could be removed and replaced by a new deeper tunnel. The bridge never returned to Washington Street, however; on August 25, 1911, construction of today's Washington Street Bridge began instead.

| First Washington Street Bridge | | | | | |
|---|---|---|---|---|---|
| Opened | Bridge type | Designed by | Constructed by | Cost | Status |
| 1891 | Swing, iron, electric powered | American Bridge Co. | American Bridge Co. (superstructure), FitzSimons and Connell Co. (substructure) | $67,594 | Transferred to Slip A at 22nd and Ashland Avenue, Jan. 3, 1907 |

# MADISON STREET BRIDGES

LOCATION: 378 West, 50 South; Madison Street runs east and west, crossing the South Branch two miles from the river mouth at Lake Michigan.

Initially designed in 1914, this bridge was part of the new Union Station plans. Its construction was delayed as City engineers decided to first complete the new Jackson Boulevard, Lake, and Monroe Street bridges before replacing the Madison Street Bridge. During the delay, the rail-height truss was introduced, and the plans for this bridge were redrawn in 1917 to include this new innovation. Like the other bridges serving Union Station that were affected by the railroad tracks on the west side of the river, the Madison Street Bridge is asymmetrical and has a short west-side counterweight and foundation. The bridge was finally constructed and opened to traffic in 1922.

In 1958 this bridge received a general rehabilitation to its superstructure and floor system, including installation of a new steel open-grid roadway. It was converted to one-person operation in 1962, and the bridge received a second major rehabilitation in 1994.

The first Madison Street Bridge was a pontoon-float swing bridge, one of five in Chicago in 1847. The bridge had semifloat-ing aprons connected to the floating draw in the center. A sketch and detailed description of this design is presented in the Introduction. This bridge and the other four Chicago pontoon swing bridges were destroyed in the Great Flood of 1849, as recounted in the sidebar in the introduction.

A second, more substantial, pontoon-float swing bridge replaced the first bridge at this location. Unable to find a contractor that would accept bonds as payment, the City appointed a committee to oversee its construction. This was the first time the City built a bridge and acted as the contractor, furnishing the materials and hiring and supervising the men to build it. This bridge was constructed of oak and pine fastened with wrought-iron fittings. It incorporated a cast-iron geared turntable that opened and closed the draw. The bridge was a substantial improvement over the last, being more durable and operating more quickly and easily with the adaptation of the recently invented railroad turntable. During the winter of 1856, this second Madison Street Bridge was removed to allow for construction of a new bridge.

Like most bridges of its day, the third Madison Street Bridge was built during the winter when there was little to no ship traffic. A cofferdam was placed in the center of the stream to build the foundation of this center-pier swing bridge, within which the piling and masonry center pier were constructed. In February

| Current (sixth) Madison Street Bridge | | | | | |
| --- | --- | --- | --- | --- | --- |
| Opened | Bridge type | Designed by | Constructed by | Cost | Status |
| Nov. 29, 1922 | Chicago type, double-leaf bascule | City Bureau of Engineering, City of Chicago, and architect Edward Bennett | Ketler-Elliott Co. (superstructure), FitzSimons and Connell Co. (substructure) | $1,622,000 | Currently in use |

Looking north at the current Madison Street Bridge during construction to shore up the western bridge house of the Monroe Street Bridge in 2010. © Kevin Keeley

1857, a flood similar to that in 1849 sunk a large pile-driving barge at Madison Street. The wreckage was forced up against the cofferdam and, along with winter ice and debris, created a dam. The waters rose, flooding the surrounding lumberyards along the river and covering much of Market Street (now Lower Wacker Drive). The mounting pressure shifted the cofferdam and damaged some of the work inside. Afterward, construction was resumed, and the substructure was completed. Political posturing by businessmen and landowners, wrangling over assessments for raising the street grade and curb improvements, caused delays. The bridge was finally completed in the fall of 1857. The street and sidewalk improvements were treated as a separate issue and completed through various means over the next two years.

This third Madison Street Bridge combined two Moseley tubular iron-arch bridges. The first tubular iron-arch bridge in the United States was constructed in 1854 near Cincinnati by

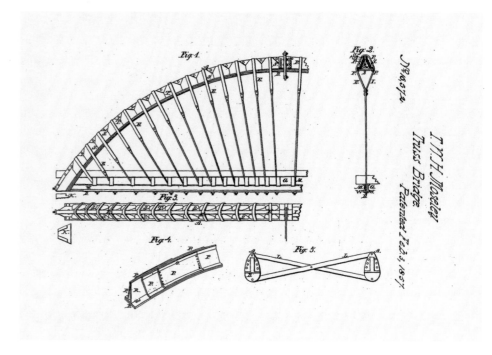

Patent drawings of a Thomas Moseley tubular iron truss, patented in 1857, used on the third Madison Street Bridge.

| First, second, and third Madison Street Bridges | | | | | |
| --- | --- | --- | --- | --- | --- |
| Opened | Bridge type | Designed by | Constructed by | Cost | Status |
| 1847 | Pontoon swing, wood, hand operated | Built to plans furnished by the Common Council | S. Peck, L. M. Boyce, and E. S. Wadsworth | $3,000 | Destroyed by flood on Mar. 12, 1849 |
| June 1849 | Pontoon turntable, wood, hand operated | Common Council, City of Chicago | Common Council, City of Chicago | Est. $4,000 | Removed in winter of 1856 |
| June 1857 | Swing, tubular iron arches, hand operated | Gregory, Bandon & Robinson using two Moseley tubular iron-arch bridges | Gaylord | $42,000 | Moved to Archer Avenue in 1875 |

Thomas W. H. Moseley and patented in February 1857.[1] The contractors for the City of Chicago, Gregory, Bandon, and Robinson, placed these two iron-arch bridges end to end over the center pier and the turntable to create the 155-foot Madison Street Bridge. It offered 62-foot draws on either side of the center pier and included a center tower with 2½-inch iron support rods extending from the tower to the far ends of each iron arch.

The third Madison Street Bridge was untouched by the Great Fire and survived another four years before being removed in 1875 to make room for a new bridge.

The old center pier was reused, while the pile and masonry approaches and sidewalks were rebuilt to accommodate the new 158' × 30½', 132-ton bridge. The fourth Madison Street Bridge was a Pratt-truss superstructure 2 feet longer and significantly wider than its predecessor. Consequently, the protection pier was widened, and piling clumps were added to help fend off ships. Reuse of bridge substructures provided a significant cost savings and had become common practice, as wood super-structures typically needed replacement after nine to ten years, whereas piling and masonry foundations were often serviceable for twenty or thirty years. The new iron bridge was only the second bridge in Chicago to use steam power (the first being the Randolph Street Bridge, which opened a year earlier). A new brick bridge house was built, and the old one was moved to Archer Avenue. Unlike today, these bridge houses were situated on the riverbank instead of being part of the bridge structure. In 1879 the bridge received a new floor and floor joists. Though this all-iron superstructure was replaced by a larger bridge after sixteen years, it was moved to Washington Street and used for another sixteen years.

The fifth Madison Street Bridge was built at the very end of the swing-bridge era and was the first two-truss, four-lane bridge in Chicago. Previous four-track bridges used three-truss super-structures. Designed to carry up to four wagon teams running abreast, the ends of this exceptionally wide Pratt-truss bridge just cleared the buildings on both sides of the river when it ro-

## Fourth and fifth Madison Street Bridges

| Opened | Bridge type | Designed by | Constructed by | Cost | Status |
|---|---|---|---|---|---|
| 1875 | Swing, iron, steam powered | American Bridge Co. | American Bridge Co. | $15,000 | Moved to Washington Street in 1891 |
| Oct. 16, 1891 | Swing, steel, electric powered | Unknown | Riter & Conley (superstructure), FitzSimons and Connell Co. (substructure) | $104,500 | Removed Sept.–Nov. 1922 |

tated. Decorated with elaborate ornamentation, this sturdy steel bridge also carried the heaviest streetcars, which shared one lane in either direction with street traffic. After two years of heavy traffic, the turntable support girders were showing some very ominous fractures, and instrumental observations determined the girders of the superstructure to be gradually collapsing. In June 1893, steel angles were riveted to the superstructure to stiffen the points of fracture. By the following year, the turntable rollers were so worn that the pinion bracket was wearing on the circular rack. The situation was remedied by planing three-eighths of an inch off the top of the rack, as had been done at Clark Street. In 1896 this bridge received a new floor system, and the following year new cast-steel rollers were installed using the interlocking-wedge system first tried at Clark Street. Twenty-one years later, in 1922, this swing bridge was removed to make way for a new bascule bridge.

Looking northeast at the fifth Madison Street Bridge. Courtesy of David R. Phillips.

# MONROE STREET BRIDGE

LOCATION: 378 West, 50 South; Monroe Street runs east and west and crosses the South Branch two miles from the river mouth at Lake Michigan.

HISTORICAL HIGHLIGHT: The Monroe Street Bridge is the only existing Chicago highway bridge entirely paid for by private funds.

The current bridge at Monroe Street is the first at this location and provides a clear river channel of 165½ feet. Four railroads combined to form the Union Station Company and create a new central railway station to represent Chicago's status as America's railway hub. The railroads wanted the station to make an architectural impact similar to that made by Union Station in Washington, D.C., which opened in 1908, and New York's Grand Central Terminal, which opened in 1913. The Union Station Company agreed to pay for this bridge in 1914 as part of negotiations with the City of Chicago to determine the site of the new station. This concession helped secure today's Union Station site on the west bank of the river between Jackson Boulevard and Adams Street, instead of at the City's proposed 12th Street site.

A crossing at Monroe Street was part of Burnham and Bennett's *Plan of Chicago* to make Monroe Street a major thoroughfare and help relieve congestion in the Loop. In so doing, the street connected the West Side to Grant Park and eventually Lake Shore Drive. The bridge provided the additional benefit of carrying traffic from Jackson Boulevard, Adams, Madison, and Lake streets during construction of the new bascule bridges and the removal of the antiquated center-pier swing bridges.

The Monroe Street Bridge, nearly identical to the Franklin-Orleans Street Bridge, is a second-generation Chicago-type bascule bridge. The plans for both bridges were drawn up based on the standardized Chicago type and were modified to meet the many requirements and challenges of the Monroe Street site. Railroad tracks along the river's west bank allowed very little room for the counterweight pit, and several underground obstacles also complicated the design. These included an abandoned water tunnel that ran diagonally beneath the riverbed at Monroe Street and two underground freight tunnels that ran under the east riverbank. As a result, the bridge was built with asymmetrical leaves. On the west bank, a shorter arm and heavier cast-iron counterweight were used, while a longer, standard, and less costly concrete counterweighted arm was used on the east bank.

| Monroe Street Bridge | | | | | |
|---|---|---|---|---|---|
| Opened | Bridge type | Designed by | Constructed by | Cost | Status |
| Feb. 22, 1919 | Chicago type, double-leaf bascule | City Bureau of Engineering and architect Edward Bennett | Ketler-Elliott Co. (superstructure), FitzSimons and Connell Co. (substructure) | $525,447 | Currently in use |

Looking north at the Monroe Street Bridge in 2009. © Kevin Keeley

Both leaves were supported by concrete-pier foundations that reached 117 feet below city datum to bedrock. (*Datum* is a surveying term that refers to a point of reference or plane for measuring elevations. In Chicago, this point of reference is specifically called out in the Municipal Code. Chicago's city datum is a plane located 17.640 feet below a mark in the bottom stone of granite of the Northern Trust Company Bank Building at La Salle and Monroe streets. This mark is indicated with a metal plaque on the southeast corner of the building.)

Originally, both the Franklin-Orleans and Monroe Street bridges had decorative terra-cotta pylons at the four corners of the bridge. They contained warning lights to stop traffic during bridge openings. In 1932 the badly deteriorated pylons were removed and replaced with more modern signals, and two new electric gates were installed at the west approach. The Monroe Street Bridge received a new roadway and sidewalks in 1947.

In 1956 the Monroe Street Bridge was converted to one-person operation. This was only the second bridge in Chicago to be converted, and a brand-new technology was installed: a closed-circuit television system that would enable the bridge tender to remotely monitor pedestrian and vehicular traffic on the far shore. This made the original western bridge house operationally superfluous, yet for many years it was preserved as part of the bridge's architecture. In 2002 the east bridge house was removed and replaced with a chintzy plywood structure. Then, in 2007, the City began a $4.3 million project, replacing both bridge houses according to the specifications of the original plans. This fourteen-month reconstruction included electrical and mechanical upgrades, new granite bases, interior wood paneling, warning bells, terra-cotta exteriors, and architectural ornamentation. Paid for with state and federal funds, this project was completed in July 2008. In 2010 it was discovered that the west bridge house was sinking, and additional work adding new supporting piers was undertaken that summer to correct the problem.

# ADAMS STREET BRIDGES

LOCATION: 380 West, 300 South; Adams Street runs east and west, crossing the South Branch 2.1 miles west of the river mouth at Lake Michigan.

Started in 1923, construction of the Adams Street Bridge was halted and plans altered to accommodate construction of Chicago's Union Station. Opened in 1925, the new train station and connecting train tracks required that the bridge be raised and the west leaf shortened. Like its neighbors at Monroe Street and Jackson Boulevard, the Adams Street Bridge has nonstandard, short west leaves featuring compact cast-iron counterweights, narrow foundations, and small counterweight pits needed to accommodate the rail lines.

Heavily influenced by the Alexandre III Bridge over the Seine in Paris, the fourth and current Adams Street Bridge well expresses the ideal Chicago-type bascule bridge. Its low arch and below-deck trusses present a sleek profile that fits well with the proportional stonework of the approaches and bridge houses. This was the City's first deck-truss drawbridge, although the very first deck truss in Chicago was located at Jackson Boulevard. A *deck truss* means that the structural steel trusswork is placed and supports the roadway entirely from below. Mayor "Big" Bill

Thompson cut the ceremonial yellow ribbon to open this 265' long by 95' wide span.

In 1956 the Adams Street Bridge was converted to one-person operation. The latest control systems, as well as new safety and warning signals, were installed to allow operation of this double-leaf bridge from a single bridge house. The bridge was redecked with an open-grid steel roadway; it also received new sidewalks and refurbished mechanical equipment in 1969. In 1996 this bridge received a complete rehabilitation by the City.

The first Adams Street Bridge was a good example of how difficult it could be for early residents to get a new river crossing. As early as 1857, after an order to advertise for bids on a second Van Buren Street Bridge, it was suggested that bids for an Adams Street bridge also be advertised. The idea was that, with the recent widening of the South Branch, the river was now the same width at both spots, and so the bridges could be virtually identical. The suggestion was rejected, though, and in 1864 a local bridge committee formed to petition the Common Council and gather subscriptions for an Adams Street Bridge. Two years later, the *Chicago Tribune* reported that "the most pressing want just now is for a bridge at Adams Street, to serve the double purpose of saving time, and distance . . . and relieving . . . the great army of teams and carriages[.] The bridge is wanted now, and every

| Current (fourth) Adams Street Bridge | | | | | |
|---|---|---|---|---|---|
| Opened | Bridge type | Designed by | Constructed by | Cost | Status |
| Aug. 26, 1927 | Chicago type, double-leaf bascule | City engineers Thomas Pihlfeldt and Donald Becker and architect Edward Bennett | Strobel Steel Construction Co. (superstructure), FitzSimons and Connell Co. (substructure) | $2.5 million | Currently in use |

Looking north at the current Adams Street Bridge in 2009. © Kevin Keeley

## First Adams Street Bridge

| Opened | Bridge type | Designed by | Constructed by | Cost | Status |
|--------|-------------|-------------|----------------|------|--------|
| 1869 | Swing, wood and iron, hand operated | Fox & Howard | Fox & Howard | $36,860 | Destroyed by fire, Oct. 8–10, 1871 |

month that it is delayed tens of thousands of dollars are lost to the citizens of Chicago, in the destruction of time which, to the businessman, is always money."[1] Adams Street was situated in the middle of a four-block stretch of river between the Madison and Van Buren Street bridges.

Political wrangling complicated the effort. The Pittsburgh, Fort Wayne & Chicago Railroad offered twenty-five thousand dollars for a tunnel at Adams Street or twenty thousand dollars for a tunnel at Washington Street. In exchange, the railroad wanted the City to abandon Adams Street from the river west to Canal Street and allow construction of a railway depot and station. The Adams Street Bridge Committee enlisted the City's best legal counsel to block the efforts of the railroad and maintain the Adams Street right-of-way. A five-year battle ensued, but the first Adams Street Bridge was built and opened in 1869. Although this wood and iron swing bridge was destroyed two years later in the Great Fire, the right-of-way on Adams Street was preserved, and there were no challenges to a second bridge.

The second swing bridge at Adams Street opened in 1872. This iron bridge was nearly identical in dimensions to the first, at 32 feet wide and 160 feet long. It served for seventeen years until it became inadequate for the busy crossing and hindered navigation. The iron structure was removed, floated via scow downriver, and reused at Taylor Street for another ten years.

Looking south at the second Adams Street Bridge in the 1870s or 1880s.

The third Adams Street Bridge was a massive steel Pratt-truss swing bridge 259 feet in length and, at 59 feet, nearly twice as wide as its predecessor. This four-track bridge greatly improved both street traffic and river navigation, as dredging and widening of the river at this crossing provided an additional

## Second Adams Street Bridge

| Opened | Bridge type | Designed by | Constructed by | Cost | Status |
|--------|-------------|-------------|----------------|------|--------|
| 1872 | Swing, iron, hand operated | Keystone Bridge Co. | Keystone Bridge Co. (superstructure), Fox & Howard (substructure) | $47,790 | Moved to Taylor Street in 1889 |

Aerial view northwest of the third Adams Street Bridge in 1923. Courtesy of CDOT.

## Third Adams Street Bridge

| Opened | Bridge type | Designed by | Constructed by | Cost | Status |
|--------|-------------|-------------|----------------|------|--------|
| 1889 | Swing, steel, electric powered | Keystone Bridge Co. | Keystone Bridge Co. (superstructure), Fox & Howard (substructure) | $141,115 | Closed to traffic, Aug. 31, 1925, and removed, Sept. 1925– Jan. 1926 |

50 feet to each draw. Interestingly, the west approach was 2¼ inches lower than the east approach, and the bridge could rotate in both directions and still match the abutments. The turntable was cleverly set at a slight angle, using a 0.87 percent grade. Construction of this bridge was completed in six months, and the cost was split between the City and the Chicago Passenger Railway Company (which ran its streetcars over the bridge).

In 1895 maintenance of this bridge included replacing the eighty cast-iron rollers with forty cast-steel rollers. On October 21, 1898, the bridge opened for the first time after conversion from steam to electric power. The bridge was repainted with two coats of red lead and oil paint below the bridge deck and one coat of black carbon paint over the rest of the superstructure in 1900. The subplanking, pavement, and sidewalks were also entirely replaced. A new protection pier was built in 1909, which work included driving 191 new pilings, and the old bridge house was removed to allow for the operation of larger, improved streetcars. A new bridge house was built over the sidewalk.

In 1917, however, the bridge was declared an obstruction to navigation by the U.S. Army Corps of Engineers. The City postponed removal as long as possible, citing a lack of funds to undertake wholesale replacement of the numerous swing bridges that had been declared obstructions by the Corps. On August 31, 1925, this bridge was finally closed to traffic. The superstructure was removed, and on October 23 any masonry on the center pier located above water was loaded directly onto scows. Dynamite was used to shatter the rest of the old center pier, and the rubble was dredged from the river to complete the removal.

# JACKSON BOULEVARD BRIDGES

LOCATION: 375 West, 300 South; Jackson Boulevard runs east and west, crossing the South Branch 2.2 miles from the river mouth at Lake Michigan.

HISTORICAL HIGHLIGHT: This Strauss design was the first deck-truss bascule highway bridge in Chicago.

In January 1911, a new Strauss bascule bridge was constructed after the removal of the first Jackson Boulevard Bridge was ordered by the secretary of war. The Sanitary District built this below-deck truss bridge for the City of Chicago. It is the only Strauss highway bridge in Chicago and selected as the only design capable of meeting the space requirements given the railroad lines along the west bank of the river. The bridge is just over 202 feet from trunnion to trunnion, offers a 173½-foot-wide opening for ships, and its 64-foot-wide highway bears four lanes of traffic with sidewalks on either side.

On March 15, 1937, the Jackson Boulevard Bridge was the scene of a terrible accident. At three o'clock that morning, the sand barge *H. Dahlke* signaled for the bridge to open, but the west leaf would not rise. It was later determined that a blown fuse had disabled the west-barricade gates, warning bells, and lights, all of which had to be active before the bridge motors could be engaged. Conferring by phone and with no traffic in sight, the two bridge tenders decided to override the safety interlock, thereby preventing the east leaf of the bridge from opening. Their concern was the approaching *Dahlke,* the crew of which was unaware of the problem. The east leaf was raised to allow the barge to pass when a car driven by Livio Cortesia, twenty-one years of age, approached from the west. The bridge tender, recognizing the danger, grabbed a red lantern and began yelling and signaling out of the second-floor window of the west bridge house. The signal went unnoticed, and Cortesia continued onto the bridge, which was slippery from a light snowfall earlier in the evening. Cortesia was able to brake, but too late. His vehicle plunged into the river, killing him. The coroner's inquest ruled it an accidental death, but recommended that "the city install hand controlled emergency gates on all bridges."[1] To date, this has not been done, but there have also not been any reported cases since this incident of a bridge opening without safety signals.

The next year, this bridge was redecked, the old sidewalks replaced and paved, and the center-lock machinery refurbished and moved under the roadway. In 1962 this bridge was converted to one-person operation.

## Current (second) Jackson Boulevard Bridge

| Opened | Bridge type | Designed by | Constructed by | Cost | Status |
|--------|-------------|-------------|----------------|------|--------|
| Jan. 29, 1916 | Strauss, trunnion bascule | Strauss Bascule Bridge Co. | Strobel Steel Construction Co. (superstructure), Great Lakes Dredge & Dock Co. (substructure) | Unknown | Currently in use |

Looking southeast at the current Jackson Street Bridge in 2009. © Kevin Keeley

Looking north at the first Jackson Boulevard Bridge in 1914. Note that the protection pier has been removed. Courtesy of MWRD.

The first Jackson Boulevard Bridge was a three-Pratt-truss steel swing span 280 feet long and 58 feet wide. Its center pier was positioned near the west bank of the river, resulting in an east arm that spanned most of the South Branch, while the west arm spanned a small portion of the river, the tracks of the Pennsylvania Railroad, and a covered bypass built over the water to carry wagon traffic under the bridge.

The construction of this first Jackson Boulevard Bridge in 1888 and the second Dearborn Street Bridge the same year ended the practice of building bridges only at alternate streets. The propeller ships on the Great Lakes were longer than some city blocks, so it was reasoned that having bridges too near one another in an already crowded river could be dangerous and force bridges to remain open significantly longer, further

## First Jackson Boulevard Bridge

| Opened | Bridge type | Designed by | Constructed by | Cost | Status |
|---|---|---|---|---|---|
| Aug. 25, 1888 | Swing, steel, steam, then electric powered | Unknown | Detroit Bridge Co. (superstructure), FitzSimons and Connell Co. (substructure) | $131,633 | Removed in 1915 |

interfering with travel. This problem was particular to the swing-bridge era, as operation of these bridges on a horizontal plane required a large swing radius in order to open and close. Because a large, open space was necessary, bridges in close proximity to one another would necessitate the simultaneous opening of bridges and greatly limit docking in the river.

In 1895 the bridge received forty new cast-steel rollers to replace the eighty original cast-iron rollers. Two years later, it was converted from steam to electric power by the City Department of Bridges. In 1899 this bridge was entirely replanked and repaved; all ironwork below the flooring was scraped and repainted, and the entire structure was thoroughly overhauled. Again replanked and repaved in 1906, it also received new sidewalks and installation of a new danger-signal system at the approaches, similar to the one in use at Rush Street. In 1910 the bridge was again replanked and, the following year, repaved. That year a new gear shaft was installed, and the end rollers, locks, and spider rods were repaired, just before the bridge was declared an obstruction to navigation and ordered removed. It was replaced with the new bascule bridge in 1915.

# VAN BUREN STREET BRIDGES AND TUNNEL

LOCATION: 361 West, 380 South; Van Buren Street runs east and west, crossing the South Branch 2.3 miles from the river mouth at Lake Michigan.

HISTORICAL HIGHLIGHT: Van Buren Street, and the Metropolitan elevated-train bridge that operated a half block south, opened in 1895 and were the very first Scherzer rolling-lift bridges in the world.

This sixth Van Buren Street Bridge was a Chicago-type bascule and widened the crossing from two to four lanes. This project followed on the heels of three major roadway projects surrounding Van Buren Street—the new Congress Street Bridge, the Eisenhower Expressway, and the extension of Upper and Lower Wacker Drive south along the line of old Market Street from Lake Street to Congress Expressway.

City engineers from the Division of Bridges and Viaducts, directed by chief bridge engineer Stephen J. Michuda, completed the design, contract, plans, specification, and supervision of the bridge's construction. The design is distinguished by having only one bridge house for a double-leaf span. Earlier Chicago-type bascule bridges required two bridge houses, one for each bridge leaf. This was only the third double-leaf bascule designed and constructed for one-person operation using a single bridge house, the first two being at State Street and at North Halsted Street.

The first Van Buren Street Bridge was a wood pontoon bridge built from the same plans as the second Madison Street Bridge, with a turntable and raised float added to put the bridge even with street level. During its seventh year of use, this bridge was severely tested by a flood on February 8, 1857. An immense choke point formed out of ice and debris that threatened to destroy the bridge and flood the surrounding area. The bridge was swung open, but for several hours it was feared the crush would overwhelm the Madison Street Bridge and precipitate "an immense destruction of property and loss of a number of lives."[1] Fortunately, the ice was released in small quantities and floated downriver without any serious damage. This Van Buren Street Bridge was removed in 1858 and replaced by a new swing bridge.

The contract for the second Van Buren Street Bridge was awarded to the newly formed firm of Chapin & Howard, which underbid all competitors and promised to construct the bridge in ninety days. As the firm's first contract, many believed this deadline was impossible to meet, yet Chapin & Howard completed the bridge on time and without sacrificing workmanship

| Current (sixth) Van Buren Street Bridge | | | | | |
|---|---|---|---|---|---|
| Opened | Bridge type | Designed by | Constructed by | Cost | Status |
| Dec. 5, 1956 | Chicago type, double-leaf bascule | Division of Bridges and Viaducts, City of Chicago | Overland Construction Co. (superstructure), M. J. Boyle & Co. (substructure) | $3.2 million | Currently in use |

Looking north at the current Van Buren Street Bridge in 2009. © Kevin Keeley

## First and second Van Buren Street Bridges

| Opened | Bridge type | Designed by | Constructed by | Cost | Status |
|---|---|---|---|---|---|
| Jan. 1850 | Pontoon swing, wood, hand operated | Derastus Harper, city superintendent of public works | Derastus Harper, city superintendent of public works | Unknown | Removed in 1858 |
| Sept. 1858 | Pivot, wood, hand operated | Chapin & Howard | Chapin & Howard | $18,000 | Removed in 1867 |

or quality. This drew wide attention and made them prominent competitors for important structural contracts. Cancellation of a large contract by the State of Alabama, citing the outbreak of the Civil War as its excuse, brought about the firm's demise; upon returning to Chicago, William Howard left the firm to partner with Harry Fox.

This second Van Buren Street Bridge was a pivot-type bridge built from the same plans as the Randolph Street Bridge constructed two years earlier. The structure was raised above the water to allow for the passage of tugs and canal barges without opening. It connected the southwestern suburbs with the heart of the city. After nine years it was replaced.

The third Van Buren Street Bridge was an improved Howe-truss center-pier swing bridge, 163 feet long and 34 feet wide, which incorporated the truss and turntable improvements patented by J. K. Thompson. This combination wood and iron truss bridge met an early demise during the Great Chicago Fire; however, the stone center pier (as shown) survived the fire.

The fourth and final swing bridge constructed at Van Buren Street reused the old stone center pier and received a new rim-bearing turntable and hand-operated bridge superstructure

Remains of the third Van Buren Street Bridge after the Great Fire in 1871.

## Third and fourth Van Buren Street Bridges

| Opened | Bridge type | Designed by | Constructed by | Cost | Status |
|---|---|---|---|---|---|
| 1867 | Swing, wood and iron, hand operated | Fox & Howard | Fox & Howard | $18,270 | Destroyed by fire, Oct. 8–10, 1871 |
| Jan. 12, 1872 | Swing, wood and iron, hand operated | E. Sweet Jr. & Co. | E. Sweet Jr. & Co. | $13,200 | Removed on Jan. 4, 1894 |

that was of the same size and design as the bridge it replaced. This fourth bridge served for twenty-two years before it was closed to traffic on January 4, 1894, and removed. A new bascule bridge replaced it.

The Van Buren Street Tunnel was designed by city engineer Samuel G. Artingstall in 1888 for the West Chicago Street Railway Company, which would own the tunnel. It was 30 feet wide and 1,514 feet long, with two 602-foot approaches consuming most of the tunnel's length. Construction started in 1890 just north of the Van Buren Street Bridge, where a watertight cofferdam extending half of the river's width was installed. This allowed for ship traffic while workers inside the rectangular dam excavated the river bottom and built the tunnel. Made mostly from Bedford limestone, the top of the masonry structure would allow a river depth of 17 feet.

In 1899, in conjunction with the reversal of the river, the Sanitary District prescribed a minimum depth of 26 feet in the river and canal. A secretary of war order dated September 27, 1904, ordered the lowering or removal of the three river tunnels to allow an increased river depth of at least 21 feet. This meant the three tunnels under the Chicago River currently allowing a maximum depth of 17 feet were now obstructions to navigation and must be reconstructed. At that time, an average of 540 streetcars carried approximately thirty-seven thousand passengers through the tunnel each day. A train would typically pass through this tunnel every two minutes during morning and evening rush hours. Each train was usually composed of three streetcars, one grip car, and two trailing cars.

A protracted legal battle between the City and the West Chicago Street Railway Company ensued over the responsibil-

## Van Buren Street Tunnel

| Opened | Tunnel type | Designed by | Constructed by | Cost | Status |
|---|---|---|---|---|---|
| 1892 | Arch masonry and concrete | Samuel G. Artingstall, civil engineer | FitzSimons and Connell Co. | $1 million | Closed in 1924 |

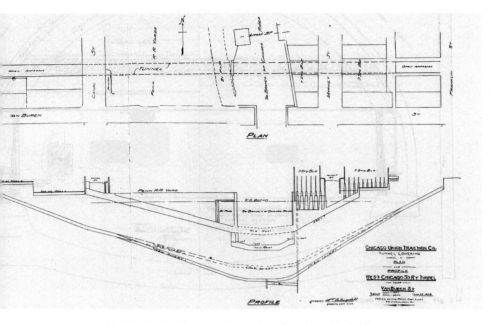

City drawings for the planned lowering of the tunnel in 1908.

through this tunnel until 1924; after that it was used only as part of a practice course for training new streetcar motormen. Many potential subway plans called for its use, but after 1952 the tunnel was leased for automobile parking and has since been either sealed or filled.

In 1895 the fifth Van Buren Street Bridge was one of only two Scherzer rolling-lift bridges in the world.[2] It was the city's third experimental design, though it had a twin built the same year by the Metropolitan West Side Elevated Train Company a half block north. This design had few moving parts, opened and closed quickly, was less expensive than competing designs, and was the most promising of the experimental designs at the time. Between 1895 and 1905, the Scherzer rolling-lift bridge was the preferred bascule bridge. By 1899, however, this design was rapidly losing favor with City engineers due to significant repair and maintenance costs.

This first Scherzer was noted in the *Department of Public Works Annual Report of 1895* as having numerous flaws. The bridge opened in February, and by May the electromechanical system had already failed, closing the bridge for four weeks. While under repair, it was determined that excess counterweights on the leaves had caused the failure. Once resolved, the bridge operated without incident throughout the summer, but by Oc-

ity of lowering this tunnel, and both refused to expend funds to comply with the order. In 1906 the U.S. Supreme Court ruled against the railway company, forcing them to pay for the tunnel's removal. By the end of that year, the Union Railway Company had acquired the West Chicago Street Railway Company and took on this construction project. A new deeper tunnel opened to trolley traffic in 1911. Regular streetcar traffic was routed

## Fifth Van Buren Street Bridge

| Opened | Bridge type | Designed by | Constructed by | Cost | Status |
|---|---|---|---|---|---|
| Feb. 4, 1895 | Scherzer rolling lift, steel, electric powered | Scherzer Rolling Lift Bridge Co. | C. L. Strobel (superstructure), FitzSimons and Connell Co. (substructure) | $169,700 | Removed in 1956 |

Looking north at the fifth Van Buren Street Bridge, with the Metropolitan Elevated Train Bridge and Jackson Boulevard Bridge in the background in 1908. Courtesy of MWRD.

tober it was clear the original electric motors were insufficient and required overhauling. The following year, the entire electrical plant was reorganized, the bridge houses were removed from the center of the roadway, and more substantial structures were constructed outside the sidewalks. This gave the bridge tenders a better view of the river and roadway. A deal was also struck in which the West Chicago Street Railway Company would supply power to the bridge, saving the City eighteen hundred dollars a year. Finally, the center lock, which had not worked for more than a year, was modified and put back into service.

In 1899 the Sanitary District dug out a 50-foot-wide channel from Van Buren Street to Adams Street. This increased the cubic flow of water needed for reversal of the river, but it also prompted considerable repair and the addition of piling clumps to protect both bridges. In 1902 two new G.E. 67 motors were installed in the west engine room, the machinery was repaired, and new centrifugal pit pumps were installed in the east engine room. The old Westinghouse motors on the east leaf were replaced in 1907 with new G.E. 58 motors, and new signal lights and semaphores were installed.

This bridge was ordered removed before April 1915 by the secretary of war due to its narrow 90-foot draw, for want of a 170-foot draw. Several extensions pushed the deadline to December 1922, but the fate of this bridge was tied to its twin, owned by the Met. A court ruled that the War Department had no authority over the Metropolitan West Side Elevated Train Company to force the removal of its bridge. The War Department therefore decided that eliminating the Van Buren Street Bridge would be futile, and the City of Chicago was given additional time. It was understood that, once the Met removed its bridge, the City would need to be ready with a new bridge. Plans for a new bascule bridge were presented and accepted by the War Department in November 1932.

In the meantime, in November 1926, George B. Dietrich, a printer by trade, was charged with disorderly conduct for racing across the Van Buren Street Bridge. As the draw was opening for a steamboat, Dietrich hit the gas, ran up the bridge, and jumped a roughly 2-foot gap, skidding down the far side and damaging his car on the abutment. This was the second known Chicago drawbridge jump *not* staged for a movie. The first is recounted in the "Michigan Avenue Bridge" chapter.

In 1931 the main trusses were reinforced, and the floor stringers and beams were replaced. A new deck was installed, low counterweights were moved up to the counterweight girders, and about forty tons of counterweights were either moved or added. In 1937 the supporting counterweight floor beams and flooring were replaced. The low center of gravity of this bridge was also improved as part of the project by adding specially designed counterweight boxes at the tail ends of the trusses. This considerably reduced stress on the operating machinery. The roadway over the counterweight blocks was rebuilt with reinforced concrete to better keep mud, dirt, and grime out of the counterweight boxes. The heel locks were also repaired. In 1946 new center-break floor beams were installed, and several other floor beams were reinforced; lateral bracing was replaced as needed, and the roadway pavement was repaired. Later that same year, a U.S. Navy scow collided with the southwest truss, and then the S.S. *Rockwood* struck the east leaf and knocked it 2 inches out of alignment. This led to the renewal of four sets of lateral bracing members, the sidewalk brackets, and additional truss repairs.

In 1955, as was common practice, a new Chicago-type bascule bridge was built in the upright position while the bridge continued to convey traffic. Though the new bridge was completed by November, use of the old Scherzer was extended to maintain traffic to Chicago's main post office during the busy holiday season. In early 1956, the old bridge was removed, and the river channel was dredged by M. J. Boyle & Company at a cost of $332,675.

# CONGRESS STREET BRIDGE

LOCATION: 320 West, 400 South; Congress Street runs east and west, crossing the South Branch 2.4 miles from the river mouth at Lake Michigan.

As part of Chicago's postwar infrastructure, the five-million-dollar twin bridges over the South Branch at Congress Street were built to connect downtown with the Congress Superhighway (later renamed the Eisenhower Expressway, or I-290). The bridges are owned and maintained by the State of Illinois, which paid for most of the construction costs (federal funds paid for the rest). These twin Chicago-type bascule bridges are operated by the City of Chicago's Department of Transportation. The two double-leaf bridges cross a bend in the South Branch and are staggered so that the south span is thirty-one feet farther east than the north span. The twin bridges are staggered or, more formally, echeloned to substantially reduce the overall length of each span. Each is 298 feet in length and provides a 43-foot-wide roadway. The four 3,515,000-pound leaves operate in unison, using a mechanical tie-in on the gear trains of the eastern and western leaves. This linkage can be disengaged, however, so that each may be operated separately as necessary for maintenance or repairs. Each bridge carries four lanes of traffic and has a bridge house on opposite sides of the crossing. The configuration is nearly identical to the twin Ohio Street bridges.

Planned in the early 1940s, the Eisenhower Expressway was originally named the Congress Expressway. The first two and a half miles between Manheim Road and First Avenue opened in December 1955; within two weeks, an eastern four-mile stretch opened from Ashland Avenue to Laramie Avenue. During the 1960s and 1970s, this expressway extended west to the junction of U.S. 20 (Lake Street), Illinois Route 64 (North Avenue), and Illinois Route 53. The expressway then followed Route 53 just beyond Illinois Route 58 before turning northwest. This was the original route for Interstate 90. Then, in 1978, the Eisenhower Expressway was renamed I-290, and a new I-90 route followed the Northwest Expressway, now referred to as the Kennedy Expressway, through the northwest side of the city past O'Hare Airport to connect with I-290 near Route 58 and continue northwest across the country.

Originally planned for rehabilitation between March 2010 and June 2011, the Congress Street Bridge underwent a thirty-three-million-dollar refurbishment that was extended due to unexpected problems and carried on into 2012. After removing

## Congress Street Bridge

| Opened | Bridge type | Designed by | Constructed by | Cost | Status |
|---|---|---|---|---|---|
| Aug. 10, 1956 | Twin Chicago type, double-leaf bascule | Division of Bridges and Viaducts, City of Chicago | Overland Construction Co. (superstructure), M. J. Boyle & Co. (substructure) | $5 million | Currently in use |

Looking north at the Congress Street Bridge in 2009. © Kevin Keeley

the old decking on one half of the bridge while the other carried street traffic, it was discovered that the existing floor beams were no longer level, complicating installation of the new bridge deck. A subsequent redesign and changes in the fabrication of decking components added to the delay. The project included minor structural repairs, new sidewalks, replacement of the bridge-house electrical systems, and a fresh coat of paint. Each half of the bridge was repaired while in the open position to allow for passage of river traffic, while its twin maintained traffic across the transportation link. As part of a joint effort between the Illinois and Chicago departments of transportation, Congress Street was rehabilitated from Michigan Avenue to Wells Street, and the Eisenhower Expressway was repaved from the Circle Interchange to Thorndale. This coincided with completion of the Wacker Drive revitalization from Randolph Street to Congress Street. The overall project plan pays particular attention throughout to improving the landscaping, curbing, sidewalks, and pedestrian access.

# HARRISON STREET BRIDGES

LOCATION: 322 West, 420 South; Harrison Street runs east and west, crossing the South Branch 2.4 miles from the river mouth at Lake Michigan.

Mayor Richard J. Daley dedicated the third Harrison Street Bridge on August 23, 1960. A Chicago-type double-leaf trunnion bascule, it provides four lanes of traffic and was designed specifically for heavy-truck use. The City undertook an extensive engineering study to ensure that this bridge was capable of carrying the heaviest truckloads serving the nation's largest U.S. postal facility. The old Downtown Post Office, built in 1921 and boasting 2.5 million square feet of floor space, sits between Harrison and Van Buren streets and is transected by the Congress Parkway. It was vacated in 1997 when a new post office with 1.5 million square feet of continuous floor space opened on the opposite side of Harrison Street. This Harrison Street Bridge also serves the new postal facility.

Petitions for a bridge at Harrison Street began in 1872 with a well-organized local improvement committee of residents, property owners, and businessmen. An early committee proclaimed that "a bridge was essential to the success of the property owners. All were aware that communication with the South Side was restricted, the district between Van Buren and Polk streets being one of the most densely populated in the city."[1] The local bridge committee, which included future mayor Carter H. Harrison, formed to petition the Common Council for funding. City plans to straighten the river between Harrison and Polk streets added to the delay, but within five years the first Harrison Street Bridge opened to traffic. This 161-ton iron swing bridge was 175 feet long and 31 feet wide. It was the last bridge constructed for the City by the American Bridge Company. The Chicago-based firm, one of the largest and best-regarded bridge builders in the nation, was dealt a fatal blow the same year. Concurrently working on the foundations of the Poughkeepsie Bridge across the Hudson River, a cofferdam failed (on Pier No. 2), and the subsequent rework, combined with an economic depression, forced the firm into bankruptcy in early 1878. Efforts to build the Poughkeepsie Railroad Bridge languished for another eight years before a new financing company was formed and new contractors hired. The bridge was opened on December 29, 1888. It should also be noted the name American Bridge Company that is still in existence today is entirely different and was founded by J. P. Morgan in 1900 with the consolidation of twenty-nine U.S. bridge and fabrication companies.

| Current (third) Harrison Street Bridge | | | | | |
| --- | --- | --- | --- | --- | --- |
| Opened | Bridge type | Designed by | Constructed by | Cost | Status |
| Aug. 23, 1960 | Chicago type, double-leaf bascule | Division of Bridges and Viaducts, City of Chicago | Overland Construction Co. (superstructure), M. J. Boyle & Co. (substructure) | $5.75 million | Currently in use |

Looking north at the current Harrison Street Bridge in 2009. © Kevin Keeley

## First Harrison Street Bridge

| Opened | Bridge type | Designed by | Constructed by | Cost | Status |
| --- | --- | --- | --- | --- | --- |
| 1877 | Swing, iron, hand operated | Unknown | American Bridge Co. | $24,875 | Removed in 1902 |

An all too common accident occurred at Harrison Street on the evening of July 11, 1883. A horse harnessed to a small laundry wagon was spooked by a gang of hoodlums loitering on the corner of Harrison and Wells streets and ran off the east approach of the open bridge. The driver, Stephen Gleck, seventeen, and passenger Antonia Theurer, twenty-three, were thrown into the water; George Gleck, twelve, and Amelia Gleck, fifteen, were also in the wagon, but jumped to safety when the horse hesitated on the verge of the open draw before plunging into the river. Captain Tom O'Neil saved Antonia Theurer by risking his own life leaping into the river from his canal boat, the *J. H. Walker.* Both were soon pulled from the water to safety. The horse was rescued two hundred feet north of the bridge, but, sadly, Stephen could not be found and was believed drowned.

This bridge was removed in 1902 by the Sanitary District and replaced by a bascule bridge.

The second Harrison Street Bridge was a Scherzer rolling-lift bridge built by the Sanitary District for the City of Chicago. Construction was delayed, to the dismay of local businessmen and residents, as plans for this bridge were complicated by the purchase of three different designs at a cost to taxpayers of more than thirty-eight thousand dollars. Plans were first bought from the Scherzer Rolling Lift Bridge Company and then from the Hall Bascule Bridge Company, and, finally, a revised Scherzer design was paid for and used. After the old bridge was removed in early 1902, and after significant planning and construction delays, a temporary pontoon swing bridge was installed in early 1904 to aid travel. The Sanitary District contracted Lydon & Drews Company to build a temporary bridge at a cost of twenty thousand dollars. It was an electrically powered pontoon swing bridge

Looking north at the second Harrison Street Bridge in 1908.

supported by a framework and turntable at the east end, while a large scow supported the west end.

Delays continued due to difficulties getting steel, but the Scherzer rolling-lift bridge was finally completed in 1905, and the temporary bridge was removed. The new 182'-long × 51'-wide double-leaf bascule may have been the first and one of very few Scherzer deck-truss bridges. All of the supporting steelwork

| Second Harrison Street Bridge | | | | | |
|---|---|---|---|---|---|
| Opened | Bridge type | Designed by | Constructed by | Cost | Status |
| 1905 | Scherzer rolling lift, steel, electric powered | Scherzer Rolling Lift Bridge Co. | Jackson & Corbett Co. (superstructure), Lydon & Drews Co. (substructure) | $272,606 | Removed in Dec. 1959 |

below the bridge deck gave the second Harrison Street Bridge a sleek profile, but the overall clearance and low steel over the river required it to open for most ship traffic. When finished it was handed over to the City of Chicago.

Several years later, this new bridge was the scene of an assault, robbery, and urban adventure for farmer Frank Washburne. While visiting Chicago and staying with friends on the West Side, Washburne was discovered on the morning of September 18, 1912, in the counterweight pit of the Harrison Street Bridge. He had sustained cuts about the head and face and explained that two ruffians had robbed him at gunpoint. He was beaten and thrown into the river after giving up fifty dollars and a gold watch. He swam to the bridge foundation and climbed onto it, but slipped, fell into the counterweight pit, and was knocked unconscious. Upon reviving, he found himself trapped and shouted himself nearly hoarse until he was finally discovered by the bridge tender and rescued. Not wholly defeated, Washburne told his rescuers that he still had forty-five dollars in his shoe that the thugs had failed to steal.

This early bascule bridge was removed in December 1959 to allow for completion of a new Chicago-type bascule bridge.

# POLK STREET BRIDGES

(No Current Bridge)

LOCATION: 800 South, 300 West. Today there is no bridge at Polk Street, which runs east and west; this location held five successive bridges crossing the South Branch 2.6 miles from the river mouth.

This first Polk Street Bridge was a hand-me-down. In 1854 the Clark Street pontoon turntable bridge was wrecked by a steamer, and the salvaged remains were reconstructed at Polk Street in 1855. The expense of establishing and tending this new river crossing was entirely borne by residents. This was the first known occurrence of what would become a common practice of reusing bridges, usually at new, less trafficked river crossings. This rebuilt bridge lasted little more than a year before Polk Street required a new pontoon swing bridge. The second Polk Street Bridge served for twelve years before being considered "a relic of the olden time"; the "poor old thing" was removed and replaced.[1]

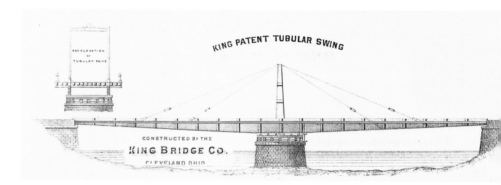

Drawing of a King Tubular Iron Bridge from an 1874 company catalog. Courtesy of Ohio Historical Society.

The third Polk Street Bridge was a combination wood and iron–truss center-pier swing bridge, 154 feet long and 31 feet wide. It was in use for two years and was destroyed by the Great Chicago Fire of 1871.

The fourth Polk Street Bridge was a tubular iron swing bridge the same length and width as its predecessor and built on the old foundation. This patented King Iron Bridge Company span was unusual, using two tubular iron trusses to create the main

| First, second, and third Polk Street Bridges | | | | | |
| --- | --- | --- | --- | --- | --- |
| Opened | Bridge type | Designed by | Installed by | Cost | Status |
| 1855 | Pontoon turntable swing, wood, hand operated | John Censor (Old Clark Street Bridge) | City superintendent of streets | Unknown | Removed in 1856 |
| 1856–57 | Pontoon turntable swing, wood, hand operated | Unknown | Unknown | $5,000 | Removed in 1869 |
| 1869 | Swing, wood and iron, hand operated | Fox & Howard | Fox & Howard | $36,595 | Destroyed by fire, Oct. 8–10, 1871 |

span, which was additionally supported by connecting rods and cables from the center towers. In addition to the drawing, the outline of this bridge is shown in the background of the image of the second Taylor Street Bridge.

In 1888 the Wisconsin Central Railroad paid for and oversaw the raising of this bridge and an approaching viaduct so the street could pass freely over their tracks. The raising of the bridge was accomplished by using scows to float the superstructure clear of the center pier. House movers transferred the structure onto the roadway to reopen the waterway, while the center pier and approaches were raised an additional three feet to match the higher grade of the new viaduct. The bridge was then replaced in a similar manner and restored to service. The entire project took approximately two weeks. The elevation of the bridge and roadway was prompted by a series of ordinances requiring the railroads either to elevate their tracks or to construct viaducts to eliminate all street and railroad crossings within the City of Chicago. This reduced congestion, improved

the flow of trains and street traffic, and increased public safety throughout the city.

This fourth Polk Street Bridge was removed in April 1907. To maintain pedestrian travel at the site of the bridge, the City provided the novel solution of a ferry boat propelled by a gasoline engine. Operation of the City ferry was suspended once the new bridge opened to traffic.

The fifth Polk Street Bridge was the first and only City-built Strauss bascule bridge. The railroad tracks owned by the Chicago Great Western Railroad on the east bank would not accommodate the larger counterweight pit and foundation of the Chicago-type bascule design. The railroad threatened an infringement suit, which would prevent use of the Chicago-type bascule design and forced the City to pay a royalty fee of fourteen thousand dollars to use the patented Strauss bascule design, which could meet the necessary space requirements. The parallel link arrangement of the Strauss below-deck counterweight allowed for a much-shorter bridge foundation. Though the pay-

## Fourth and fifth Polk Street Bridges

| Opened | Bridge type | Designed by | Constructed by | Cost | Status |
|---|---|---|---|---|---|
| June 17, 1872 | Swing, iron, hand operated | Unknown | King Iron Bridge Co. (superstructure), E. Sweet Jr. & Co. (substructure) | $37,900 | Removed on Apr. 1, 1907 |
| Sept. 2, 1910 | Strauss trunnion bascule, steel, electric powered | Strauss Bascule Bridge Co. | Standard Construction Co. (superstructure), J. J. Gallery (substructure) | $195,591 | Closed Feb. 8, 1972, and then removed |

ment of patent royalties was unavoidable, it proved a sore point with taxpayers, and the City received significant public criticism for the decision. The bridge measured 193 feet from trunnion to trunnion, it had an overall width of 40 feet, and its two trusses were 25½ feet from center to center. Each bridge leaf was operated using two 40-horsepower motors. This was not the only Strauss fixed-trunnion bascule bridge in Chicago, as, faced with similar space restrictions, the Sanitary District also built a Strauss bascule at Jackson Boulevard in 1915.

By the 1970s, traffic around Polk Street had declined, and the bridge's "electrical equipment had deteriorated badly."[2] The superstructure was in desperate need of rehabilitation. In 1972 the bridge was closed, removed, and never replaced. Remnants of the old bridge foundation can still be seen along the bank of the South Branch in line with Polk Street, just north of Marina City.

Looking north at the fifth Polk Street Bridge in 1914.

# TAYLOR STREET BRIDGES

(No Current Bridge)

LOCATION: 1000 South, 300 West. Today there is no bridge at Taylor Street, which runs east and west but no longer crosses the Chicago River. Taylor Street held two successive bridges crossing the South Branch 2.5 miles from the river mouth at Lake Michigan.

Local property owners began lobbying in earnest for a bridge at Taylor Street in 1882. A new bridge was estimated at $22,000 for the substructure, consisting of a center pier, turntable, and approaches, and $16,000 for the bridge superstructure itself. In addition, it was reported by commissioner De-Witt Cregier that 21,612 square feet would have to be purchased on the east bank, at a cost of $43,000. Dredging would also be necessary, estimated at $8,000, making the total expected cost of the bridge almost $90,000.

In 1890 the Taylor Street Bridge Committee, working with the City Council, reached an alternative solution. In exchange for the right to run their streetcars across the bridge, the West Chicago Street Railway Company would pay $25,500 for the foundation.

The City Bridge Department refitted and installed the old Adams Street superstructure to create the first bridge at Taylor Street.

This new bridge and the neighboring Chicago Terminal Transfer Railway Bridge, built in 1885, were so close to one another that their fates became inexorably linked. To allow for busy ship traffic on the South Branch, the two bridges were operated in tandem and by 1898 were deemed "one of the worst obstructions in the river" by the U.S. Army Corps of Engineers.[1] The City was saved from correcting this problem by the Sanitary District, which replaced both bridges after dredging and widening the river to attain the cubic flow of water necessary to reverse the direction of the river.

The old Adams Street superstructure was floated by scow to a slip near Oakland and 22nd streets in 1899. The City wanted to use this bridge a third time at Throop Street, but the secretary of war rejected this proposal. The old structure was stored on the scows for ten months. The process of moving it to land was nearly completed when a loud crack heralded the snapping of two bridge supports, and the entire iron structure collapsed into the river. Twelve City workers jumped into the river and swam for their lives; two of them, George Checksfield, foreman, and

| First Taylor Street Bridge | | | | | |
|---|---|---|---|---|---|
| Opened | Bridge type | Designed by | Constructed by | Cost | Status |
| 1890 | Swing, iron, hand operated | Keystone Bridge Co. | Refurbished by city day labor (superstructure), Chicago Dredge & Dock Co. (substructure) | $40,380 | Floated by scow to a slip near Oakland and 22nd Street on May 11, 1899 |

## Second Taylor Street Bridge

| Opened | Bridge type | Designed by | Constructed by | Cost | Status |
|---|---|---|---|---|---|
| Feb. 1901 | Scherzer rolling lift, steel, electric powered | Scherzer Rolling Lift Bridge Co. | Chicago Bridge & Iron | $107,323 | Removed between Dec. 6, 1928, and Feb. 6, 1929 |

Gustave Sabath, were struck and temporarily pinned underwater. Luckily, neither was knocked unconscious, and both men were able to extract themselves and swim to shore. No one was killed, but after that the bridge was never reused.

The second Taylor Street Bridge was designed by the Scherzer Rolling Lift Bridge Company. The company furnished detailed plans and supervised the erection of this bridge for the Sanitary District. This bridge served Taylor Street for twenty-eight years before it was removed so the river between Polk and 18th streets could be straightened. The Great Lakes Dredge & Dock Company handled the bridge removal, and Taylor Street has not had a bridge for more than eighty years.

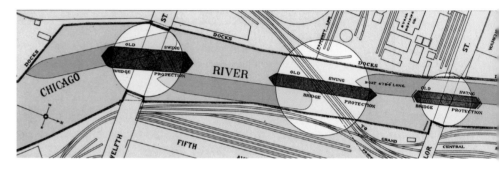

Colorized Army Corps of Engineers drawing of 12th Street, C.T.T. Railroad, and Taylor Street bridges on the South Branch in 1898. Courtesy of Dennis McClendon.

Looking north at the second Taylor Street Bridge with the fourth Polk Street Bridge in the background in 1906.

# ROOSEVELT ROAD (FORMERLY 12TH STREET) BRIDGES

LOCATION: 230 West, 1200 South; Roosevelt Road (formerly 12th Street) runs east and west, crossing the South Branch 2.9 miles from the river mouth at Lake Michigan.

HISTORICAL HIGHLIGHT: Formerly 12th Street, in July 1919, this thoroughfare was renamed Roosevelt Road in honor of President Theodore Roosevelt, who had died at the beginning of that year.

The fourth and current Roosevelt Road Bridge was a major element in the decade-long project to straighten the South Branch. Between Polk and 18th streets, a new channel was dug to the west, over which the new Roosevelt Road Bridge was erected. A series of viaducts connecting the eastern and western approaches over twenty-eight different rail lines was also built. A further challenge was that the timing had to be coordinated with the digging and dredging of the new river channel and the filling of the old, all the while maintaining an open waterway for ship traffic.

Construction on Roosevelt Road began in 1920 with the closing of the old swing bridge to street traffic. The old swing span was then removed and reinstalled as a temporary bridge while new viaducts were being constructed. Planning complications and lack of funds prevented any significant construction for six years.

Construction of the bridge over the new channel started in July 1926. In September 1928, it opened to automotive and streetcar traffic, while the temporary bridge continued to carry traffic across the old river channel. Due to grade differences between the two bridges, only the western leaf of the new bascule bridge could open for ships, but this was sufficient for the new channel to open to navigation on December 15, 1929.

The final viaduct over the old channel was opened to traffic on June 26, 1930. Since the old swing bridge was no longer necessary, it was turned over to the contractor for removal and disposal. After a temporary closure to raise the west leaf of the new bascule bridge to be level with the new road grade, permanent streetcar tracks were installed, and the entire structure was officially opened on November 22, 1930. Throughout the year, the old channel was filled using material that was hydraulically dredged from the new channel, fill from other excavations, and finally sand dredged from Lake Michigan. Elimination of the old channel was completed on December 30, 1930.

The original streetcar lines were removed in the late 1950s from the Roosevelt Road Bridge and viaducts. In 1963 City work-

| Current (fourth) Roosevelt Road Bridge | | | | | |
|---|---|---|---|---|---|
| Opened | Bridge type | Designed by | Constructed by | Cost | Status |
| Nov. 22, 1930 | Chicago type, double-leaf bascule | Division of Bridges and Viaducts, City of Chicago | Ketler-Elliott Co. (superstructure), FitzSimons and Connell Co. (substructure) | $2,050,116 | Currently in use |

Looking north at the current Roosevelt Road Bridge in 2008. © Laura Banick

ers installed a new five-inch, open steel-grid bridge deck. This bridge and viaduct system was completely refurbished in 1995.

In the area then known as 12th Street, property owners raised two thousand dollars—and received City approval and the appropriation of an additional one thousand dollars—for a bridge. Based on a plan held by the City clerk for that purpose, City street superintendent John Van Osdel constructed this first 12th Street Bridge. It opened to traffic in the fall of 1855. Though unconfirmed, it is believed that this was a substantial turntable pontoon swing bridge, as a lesser bridge would not have survived the flood of February 8, 1857. The rising ice and water piled up against the 12th Street Bridge, and it was partially swung open in an attempt to clear the channel; after many hours, however, only a portion of the ice had passed downstream. The bridge did survive, and, in 1868, this first 12th Street Bridge was removed and replaced.

In 1868 the river was widened, and a second 12th Street Bridge was built. This swing bridge was constructed under the supervision of James K. Thompson, the City superintendent of streets. It incorporated his improved turntable and refined Howe-truss design. By February 1886, however, following eighteen years of service, the 12th Street Bridge, as reported by City engineer Artingstall, was in "dangerous condition," and he recommended that it be torn down to save the City from potential

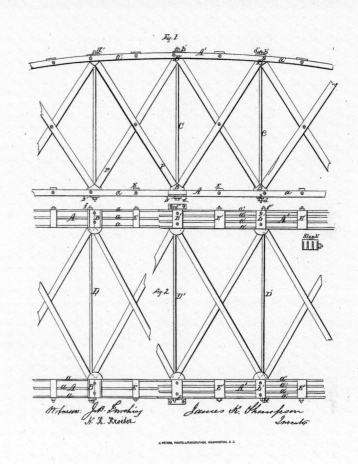

Side and top views of J. K. Thompson's truss bridge design, patented in 1868, used on the second 12th Street Bridge.

## First 12th Street Bridge

| Opened | Bridge type | Designed by | Constructed by | Cost | Status |
|---|---|---|---|---|---|
| Fall of 1855 | Pontoon swing, wood, hand operated | Unknown | City superintendent John Van Osdel | $2,877 | Removed in 1868 |

## Second 12th Street Bridge

| Opened | Bridge type | Designed by | Constructed by | Cost | Status |
|--------|-------------|-------------|----------------|------|--------|
| Apr. 28, 1868 | Swing, wood and iron, hand operated | Fox & Howard | Fox & Howard | $44,450 | Removed in 1887 |

liability.[1] In April 1887, the bridge was removed, and a new bridge was constructed that summer.

The third 12th Street Bridge was the first steel bridge built by the City of Chicago, and it was similar in design to the Rush Street Bridge built in 1884 and the Lake Street Bridge built in 1886. Chicago's very first steel bridge was built by the Chicago & North Western Railway near Kinzie Street in 1879. The Chicago Forge & Bolt Company, which had acquired the former works of the American Bridge Company at Pershing Road and Stewart Avenue next to the Pennsylvania Central rail line, built the third 12th Street Bridge. One-sixth of the center pier and the entire east abutment of this 220-foot-long, 49-foot-wide Pratt-truss bridge were paid for by the Chicago & Western Indiana Railroad Company.

The opening of the bridge, which was built in only a few months, was delayed until the connecting viaducts were completed. According to the *Public Works Annual Report of 1886,* this new crossing was "one of the most extensive structures built by the City."[2] Three viaducts extended 300 feet east of the bridge, and four additional viaducts extended 363 feet west over the multitude of railroad tracks along both banks of the river. The viaducts on the western side of the river were built at the expense of the Chicago & Great Western Railway Company, and the eastern viaducts were paid for by the Chicago Rock Island & Pacific Railroad and Lake Shore & Michigan Southern Railroad; the total cost of the project was estimated at just under $350,000.

Looking northwest at the third 12th Street Bridge in 1919, with the C.T.T. Railroad Bridge in the background. Courtesy of MWRD.

In 1889 the 12th Street Bridge sustained a blow from a passing vessel. Although it initially did not seem serious, the collision was later determined to have caused considerable damage. The stone center pier had been broken, and the bridge was shoved to the west. As a result, it was necessary to raise the bridge off the foundation so that a crew could repair the stonework and

## Third 12th Street Bridge

| Opened | Bridge type | Designed by | Constructed by | Cost | Status |
|---|---|---|---|---|---|
| June 1887 | Swing, steel, electric powered | Unknown | Chicago Forge & Bolt Co. (superstructure), FitzSimons and Connell Co. (substructure) | $101,010 | Closed to traffic in 1920; later removed and used as a temporary bridge |

install a center step and twelve spider rods to fix the turntable. The bridge was also cleaned and repainted before being returned to service, at a cost of approximately $6,000.

In 1899 the bridge was replanked and repaved. Three years later, the protection pier, boilers, machinery, floor beams, and stringers were refurbished. In 1908 a new, improved danger-signal system was installed; the end rollers, locks, spider rods, and engines were serviced; and the boilers were overhauled. In 1910 the steam plant was rebuilt, the boilers were once again overhauled, and damage to the sidewalks made by a series of col-

lisions was repaired. Temporary repairs were also made to the center pier and turntable. In 1911 the ironwork received a coat of paint, and nine clumps of thirteen pilings were added to the protection pier. The boilers were overhauled for a third time, a new crown-gear pinion shaft was installed, and the turntable components were repaired and adjusted. The disintegrating masonry of the center pier above the waterline was rebuilt in January 1912.

In 1919 the third bridge was closed to traffic, and streetcars were rerouted to another crossing. In 1920 a footbridge was substituted until the new bascule bridge was completed in 1930.

# 18TH STREET BRIDGES

LOCATION: 301 West, 1600 South; 18th Street runs east and west, crossing the South Branch 3.6 miles from the river mouth at Lake Michigan.

Although several Chicago railroad bridges are longer, the current and fifth 18th Street Bridge is the longest single-leaf bascule ever built by the City of Chicago. Spanning 182 feet, it provides a 125-foot channel for navigation when open. It was the first City bridge constructed with foreign steel: 120 tons from the Beloit-Italia Company of northern Italy. Two companies bid on the project. Over the previous decade, Overland Construction had built nearly all City bridges; following their partnership with the American Bridge Division of U.S. Steel, however, Overland was underbid by $61,000 by the Contracting & Material Company. The entire project, completed in 1967, including land acquisition, approaches, viaducts, street work, and the bridge, came to a total cost of $6.7 million. It was financed through City bridge bonds.

Originally, 18th Street was called Old Street, and the first bridge built there was a turntable pontoon swing bridge. William Linton was contracted to build this bridge based on detailed plans on file with the City clerk that were originally intended for bridges at Erie Street, Grand Avenue, Polk Street, and 18th Street. The design specified a cast-iron turntable supported by two concentric circles of pilings. The planked roadway of the bridge draw was supported by two beams connected to the turntable and supported at the other end by a sixty-four-by-twenty-one-by-seven-foot wood pontoon. The floating end of this arrangement connected with a semifloating apron to allow for changing water levels. A fixed piling and timber approach connected this structure with the street on either end. In 1867 the first 18th Street Bridge was removed and replaced by a new swing bridge.

The second 18th Street Bridge was designed and built under the supervision of street superintendent J. K. Thompson. His improved Howe-truss and turntable design was used for this bridge. In 1878 the approaches were modified, and viaducts

| Current (fifth) 18th Street Bridge | | | | | |
| --- | --- | --- | --- | --- | --- |
| Opened | Bridge type | Designed by | Constructed by | Cost | Status |
| Aug. 18, 1967 | Chicago type, single-leaf bascule | Division of Bridges and Viaducts, City of Chicago | Contracting & Material Co. | $2.3 million | Currently in use |

| First 18th Street Bridge | | | | | |
| --- | --- | --- | --- | --- | --- |
| Opened | Bridge type | Designed by | Constructed by | Cost | Status |
| 1857 | Pontoon turntable swing, wood, hand operated | Unknown | William Linton | $3,800 | Removed in 1867 |

Looking north at the current 18th Street Bridge in 2009. © Kevin Keeley

were added over the tracks of the Pittsburgh, Fort Wayne & Chicago Railroad and the Chicago, Alton & St. Louis Railroad and at Lumber Street. Awarded the contract on August 3, the Keystone Bridge Company of Pittsburgh completed iron viaducts that included a side approach at Lumber Street. The viaducts opened for travel on December 20. The *Department of Public Works Annual Report* considered it in "every respect a superior" viaduct and an ornament to the City of Chicago; it was constructed of the best-quality iron at a cost of $11,194.[1] The following year, the bridge was substantially repaired, particularly

## Second 18th Street Bridge

| Opened | Bridge type | Designed by | Constructed by | Cost | Status |
|---|---|---|---|---|---|
| 1867 | Swing, wood and iron, hand operated | Fox & Howard | Fox & Howard | $28,500 | Removed in 1888 |

its lower chords. The 18th Street Bridge served this crossing for twenty-one years, but, by 1886, the City engineer reported that the "bridge was worn out."[2] Two years later, it was replaced.

In 1888 a third 18th Street Bridge was constructed. It was operated by hand until 1894, when the steam plant removed from the Lake Street Bridge was installed. In 1896 a new floor was added, and the turntable wheels were replaced using Roemheld's interlocking-wedge system, first used at Clark Street. After fifteen years of service, this swing bridge was removed. The six-hundred-ton, 150-foot-long Pratt-truss superstructure was raised and floated on scows to the city yard near South Ashland Avenue. Prior to this move being undertaken, the dam at Lockport was temporarily raised to check the current in the waterway and aid the safe transport of the bridge down the river. Its superstructure was refurbished and installed at South Western

Looking north at the third 18th Street Bridge in 1903. Courtesy of MWRD.

## Third 18th Street Bridge

| Opened | Bridge type | Designed by | Constructed by | Cost | Status |
|---|---|---|---|---|---|
| Nov. 6, 1888 | Swing, iron and steel, hand operated | Unknown | King Bridge Co. (superstructure), Chicago Dredge & Dock Co. (substructure) | $62,788 | Moved to City yard near S. Ashland Avenue on Nov. 15, 1903 |

## Fourth 18th Street Bridge

| Opened | Bridge type | Designed by | Constructed by | Cost | Status |
|---|---|---|---|---|---|
| 1905 | Scherzer rolling lift, steel, electric powered | Scherzer Rolling Lift Bridge Co. | Jackson & Corbett Co. (superstructure), Lydon & Drews Co. (substructure) | $218,805 | Removed in Apr. 1966 |

Avenue over the West Fork of the South Branch in 1905, where it served until the 1930s. The bridge was no longer needed, as the Ogden-Wentworth Canal and West Fork of the South Branch had been filled in and thus removed from the landscape.

The fourth 18th Street Bridge was a Scherzer rolling-lift bascule built by the Sanitary District. The widening of the South Branch to as much as 200 feet as part of the river-flow reversal project meant several bridges had to be replaced. In 1900 the cash-strapped City prevailed upon the Sanitary District to build eight new bridges, including this 18th Street Bridge.

In 1938 this bridge received considerable attention. New 3-inch soleplates and 2¼-inch track plates replaced the original 2-inch sole- and track plates; new deck steel, fixed-approach spans, anchor columns, and strut-roller tracks were installed; the main truss members were reinforced; the floor breaks were replaced and a new roadway and sidewalks installed; reinforced concrete slabs were installed on the rear ends of the roadway and on the fixed approaches; cast-iron counterweights attached to the rear floor beams were reinstalled to a new platform ahead of the floor beams to better protect them from rust; the bridge was repainted; four new electric gates, electric lights, pit pumps, and pump houses were added; leaks in the counterweight pits were stopped; and a bridge house was built on the north side of the street to replace a western bridge house that had been destroyed in a fire. The new bridge-house location provided better visibility and was completed in early 1940. This 18th Street Bridge was removed in 1966 and replaced by the current bridge, a modern Chicago-type bascule.

Looking north at the open fourth 18th Street Bridge in 1908.

# CANAL STREET BRIDGES

LOCATION: 500 West, 2201 South; Canal Street runs north and south, crossing the South Branch 3.8 miles from the river mouth at Lake Michigan.

HISTORICAL HIGHLIGHT: This crossing is unique in that each of the four successive Canal Street bridges has been built from a different moveable design. The first Canal Street Bridge, built in 1890, provided a test case of the federal authority over the nation's navigable waterways that had been significantly strengthened by the Rivers and Harbors Act of 1890.

This fourth Canal Street Bridge was built to improve both traffic flow and navigation. The old Scherzer rolling-lift bridge it replaced accommodated only two lanes of traffic and offered a much narrower channel and lower river clearance. The project was started in October 1941, but the War Production Board halted work in 1943, redirecting material and resources to the war effort. Construction resumed in August 1946 and was completed in early 1948. This postwar span is one of the largest bascule bridges in Chicago. At 68 feet wide and with 254½ feet between the trunnions, this span provides a clear channel width of 170 feet for ships.

Canal Street parallels the South Branch just west of the Loop. At one time, it was the principal route for northbound and southbound truck traffic. Below 18th Street, the river bends west, and it is here that Canal Street bridges the river. This crossing is connected with the Central Manufacturing District south of the bridge, and the new four-lane bridge provided a major infrastructural improvement to the city.

Several refinements to the Chicago-type bascule design were first incorporated at Canal Street. The first use of an open steel-grate roadway with roller-type (instead of sleeve-type) bearings on the trunnions was introduced on this bridge. The open steel-grate roadway gives the bridge an advantage in that it reduces three key elements: weight, cost, and wind force. This was especially important to the design process because, in any position, but particularly when the leaves are open, the bridge must be capable of withstanding winds of up to one hundred miles per hour. Reducing wind resistance with the grate decking provides significant additional savings by allowing for the use of lighter building materials. The roller bearings are an improvement over sleeve bearings, as they further decrease friction, provide a longer life, and decrease the power necessary to operate the bridge. These bearings support approximately

| Current (fourth) Canal Street Bridge | | | | | |
|---|---|---|---|---|---|
| Opened | Bridge type | Designed by | Constructed by | Cost | Status |
| May 28, 1948 | Chicago type, double-leaf bascule | Division of Bridges and Viaducts, City of Chicago | Mount Vernon Bridge Co. (superstructure), Simpson Construction (substructure) | $1,455,000 | Currently in use |

Looking west at the current Canal Street Bridge in 2008. © Laura Banick

450 tons, but can support more than 1,600 tons at five revolutions per minute.

This bridge was originally designed for hydraulic operation; the necessary 125-horsepower constant-speed D.C. motors needed for the system, however, were unavailable at that time. What was supposed to be a temporary electric-motor power system was installed, and the hydraulic operating system has never been implemented. In 1963 this bridge was converted to one-person operation.

The first bridge at Canal Street was a swing bridge built by the City of Chicago in 1890. City alderman Dunham claimed that the Canal Street swing bridge was "put up to serve the private interests of a coal firm in the vicinity,"[1] but records show three major subscribers to the bridge: the Schoenhofen Brewing Company contributed three thousand dollars, Silver Creek and Morris Coal Company two thousand dollars, and the Conrad Seipp Brewing Company one thousand dollars. Canal Street was home to numerous grain elevators and coal and lumber businesses, making it a busy and important thoroughfare. An estimated seven thousand vessels annually passed through this point in the river. It proved a lightning rod, pitting commerce against shipping interests and testing the extent of federal au-

Plat of the proposed first bridge at Canal Street in 1890.

## First Canal Street Bridge

| Opened | Bridge type | Designed by | Constructed by | Cost | Status |
|--------|-------------|-------------|----------------|------|--------|
| 1890 | Swing Howe truss, wood and iron, hand operated | Abraham Gottlieb, civil engineer | A. Gottlieb Bridge Co. (superstructure), Chicago Dredge & Dock Co. (substructure) | $22,362 | Removed in 1892 to Belmont Avenue |

thority over the nation's waterways. Even as the bridge was being planned, shipping interests began protesting loudly, calling the structure "an outrage," further constraining navigation, placing this center-pier swing bridge before a bend in the river to the west.[2] The U.S. Army Corps of Engineers soon declared this new bridge an obstruction to navigation.

The center of a Supreme Court ruling upholding the Rivers and Harbors Act, a subsequent congressional amendment, and an attorney general opinion forced the reluctant U.S. secretary of war, Stephen B. Elkins, to order removal of this Canal Street Bridge in 1892. Constructed during Mayor DeWitt C. Cregier's administration, the bridge is said to have caused his defeat in 1891, as shipping interests were solidly against his reelection for creating this impediment to navigation. In 1892 Mayor Washburne's administration complied with the removal order, and the two-year-old bridge was raised up on scows and floated up the North Branch, where it replaced a fixed bridge at Belmont Avenue.

The second Canal Street Bridge was an extension of the first folding-lift experiment tried at Weed Street. Patented by Captain William Harman on June 5, 1888, the bridge employed two segmental leaves.[3] Each counterweighted leaf was hinged at the rear and at its midpoint so that when the operating machinery was set in motion, the midpoint hinge would rise and the two segments fold together. This Canal Street Bridge offered an eighty-foot

Looking east at the second Canal Street Bridge in 1900.
Courtesy of Chicago Maritime Museum.

## Second Canal Street Bridge

| Opened | Bridge type | Designed by | Constructed by | Cost | Status |
|--------|-------------|-------------|----------------|------|--------|
| June 1893 | Folding-lift bascule, steel, electric powered | William Harman | Shalter & Schnigau | $46,845 | Removed in 1903 |

draw for the passage of vessels and a twenty-foot-wide roadway with six-foot-wide sidewalks on either side. Located in the center of Chicago's coal and lumber district, the bridge held up well under the heavily loaded teams that frequently crossed, opening or closing in fifteen to twenty seconds. Although promised to be a vastly improved bridge over the first one constructed at Weed Street, it suffered from serious mechanical problems.

On June 2, 1894, about a hundred members of the Western Society of Engineers took an excursion down the river to inspect the vertical-lift bridge at South Halsted Street and the Canal Street folding-lift bridge. The Canal Street Bridge was opened and closed several times to a fearful scraping noise and grinding as though it were in need of oil. The bridge was deemed a "contraption" and needed as much as a fifth wheel on a wagon.[4] The 1896 *Public Works Annual Report* related that the bridge was "repeatedly out of service" due to the failure of its machinery and was "found impossible" to keep in operation "any length of time without making extensive repairs."[5] A rack-and-pinion operating mechanism was substituted the following year.

On October 26, 1900, the steamer *Chactaw* collided with the northwest corner of the protection pier, broke down a clump of piles, and struck a foundation pier, cracking it along its vertical length and twisting its steel tower. Then, on November 30, the steamer *Arthur Orr* struck the same pier, forcing the bridge out of service for the rest of the year. This would be the last folding-lift bridge ever built in Chicago, and it was replaced in 1903 with a different design. The nickname "jackknife" carried on and was often used to describe the early Scherzer rolling-lift bridges that followed. Only one other folding-lift bridge was ever built, at Holton Street in Milwaukee, Wisconsin, and operated from 1894 to 1926.

Looking northeast at the third Canal Street Bridge in 1908, with the Stewart Street Railroad Bridge in the background.

The third Canal Street Bridge was built by the Sanitary District as part of an extension of the Sanitary & Ship Canal. The narrow river cross-section at Canal Street increased the river's current, so it was necessary to widen the river. A new bridge would be required.

The Scherzer rolling-lift design was selected and constructed. In 1906 the bridge was redecked, the sidewalks were replaced, and two G.E. 58 motors replaced the original Westinghouse motors that had burned out. The column supporting the southeast bearing block was reinforced with heavy plates to remedy significant fractures in the support beam. In 1911 the river was dredged, requiring the submarine cables to be removed and relaid.

In 1912 one leaf of the bridge leaped forward fourteen inches toward the river, putting it out of service. An additional tooth was

## Third Canal Street Bridge

| Opened | Bridge type | Designed by | Constructed by | Cost | Status |
|--------|-------------|-------------|----------------|------|--------|
| 1903 | Scherzer rolling lift, steel, electric powered | Scherzer Rolling Lift Bridge Co. | Lydon & Drews Co. (substructure), American Bridge Co. (superstructure) | $128,000 | Removed in Nov. 1941 |

installed at the river end of the track girder in hopes that it would better hold the segmental girder; the ultimate solution, however, would not come until 1935 when the sole- and track plates of the bridge were replaced. This was undertaken while the street was closed by the county for widening and repaving north and south of the bridge. The new three-inch sole- and track plates were designed to provide a continuous gear mesh, to significantly improve upon the old two-and-a-half-inch plates, and keep the bridge leaves from jumping or breaking during lifts. A reinforced concrete-slab roadway was also installed on the rear of each bridge leaf, and the rest of the bridge deck was repaved with asphalt planking from the Johns-Manville Company. In 1941 this bridge was removed after nearly forty years of service.

# CERMAK ROAD BRIDGES

LOCATION: 501 West, 2200 South; Cermak Road (formerly 22nd Street) runs east and west, crossing the South Branch four miles from the river mouth at Lake Michigan.

HISTORICAL HIGHLIGHT: Cermak Road has the last remaining Scherzer rolling-lift highway bridge in Chicago. (The last remaining Scherzer rolling-lift *railroad* bridges in Chicago are a series of four alternating single-leaf bridges located just west of South Western Avenue, crossing the Sanitary & Ship Canal.) Originally known as 22nd Street, the road was renamed in honor of Democratic mayor Anton Cermak after he was shot and killed by Giuseppe Zangara on February 15, 1933, while traveling with President-Elect Franklin Roosevelt in Miami, Florida.

The Sanitary District built this bridge at the behest of the City to replace a swing bridge at the same location, and it was the tenth bridge turned over to the City of Chicago. The Sanitary District also built a cluster of four Scherzer railroad bridges that remain over the Sanitary & Ship Canal near 31st Street and Western Avenue, often referred to as the Pennsylvania Railroad "Eight-Track" Bridges.

The all-steel, thru-truss Cermak Road Bridge is 216 feet long and 60 feet wide. The superstructure is constructed with riveted gusset-plate connections, the abutments are made of reinforced concrete, and wood bridge houses are located on the northeast and southwest corners of the span. The bridge was closed April 1, 1997, so it could undergo a sixteen-million-dollar renovation. The structure was rebuilt and raised to a new river clearance of 19 feet, and it reopened October 28, 1998. Before this renovation was completed, the Cermak Road Bridge had the lowest clearance of any bridge on the Chicago River. The rehabilitation allowed the City to eliminate twenty-four-hour bridge tending at Cermak Road and reduced the average number of openings from 2,000 to an estimated 250 per year. The two-story, flat-roofed bridge houses were also completely rebuilt, with more pleasing single-story wood hip-roofed houses added during the renovation.

The plan for the first bridge at 22nd Street included widening the street and the river, and was approved by the Common Council on February 8, 1870. The plat shows removal of a portion of the northeast riverbank and the location of where the swing bridge would be placed. The dotted circle represents the extreme rotation of the bridge when swung. The protection pier extended beyond the bridge to protect the bridge from river traffic when in the open position. This proved to be the swing design's fatal flaw, as the protection and center pier took up a large portion of

| **Current (second) Cermak Road Bridge** | | | | | |
|---|---|---|---|---|---|
| Opened | Bridge type | Designed by | Constructed by | Cost | Status |
| Sept. 7, 1905 | Scherzer rolling lift, steel, electric powered | Scherzer Rolling Lift Bridge Co. | Jackson & Corbett Co. (superstructure), Great Lakes Dredge & Dock Co. (substructure) | $277,682 | Currently in use |

Looking northeast at the current Cermak Road Bridge in 2010. © Kevin Keeley

## First Cermak Road Bridge

| Opened | Bridge type | Designed by | Constructed by | Cost | Status |
|--------|-------------|-------------|----------------|------|--------|
| 1871 | Swing, wood and iron, hand operated | Fox & Howard | Fox & Howard | $26,900 | Removed in Sept. 1905 |

Looking west at the first Cermak Road Bridge in 1905. Courtesy of MWRD.

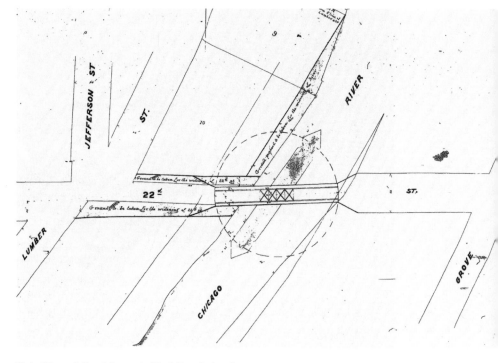

Plat of Cermak Road (formerly 22nd Street) showing proposed swing bridge and land removal in 1870.

the river channel and presented a serious obstacle to navigation. Damage caused by vessels was common, though certainly the results would have been much worse had such measures not been taken. This Howe-truss bridge was 210 feet long and 32 feet wide and rested on a masonry and pile foundation. In September 1905, this bridge was removed, and the new bridge was constructed.

# DAN RYAN (I-90/I-94) EXPRESSWAY BRIDGE

LOCATION: 700 West, 2400 South; the Dan Ryan Expressway (formerly Southwest Expressway) runs north and south through Chicago and crosses the South Branch 4.3 miles from the river mouth at Lake Michigan.

HISTORICAL HIGHLIGHT: This is the first and only fixed bridge over the navigable portion of the South Branch.

The Federal-Aid Highway Act of 1956 authorized development of the interstate highway system. The Southwest Expressway was planned; with it came the beginnings of a fixed-bridge policy for the branches of the Chicago River. In the late 1950s, the proposed fixed bridge on the South Branch stirred up a bitter controversy between maritime and highway interests. It came to a head in 1959 when the Interstate Highway Commission overseeing construction of the Southwest Expressway (later renamed the Dan Ryan) stated that they did not want a drawbridge or another huge Skyway Toll Bridge at this crossing.

Although plans for a bascule bridge were drawn up and approved by the War Department, city, state, and federal highway authorities delayed moving forward and continued pushing for a fixed bridge. They proposed a fixed bridge that would cost eight million dollars less than a bascule bridge and offer a sixty-foot river clearance. Outraged navigational interests argued that sixty feet was too low and that the fixed span would cut off access to five miles of riverbank, handicapping all masted vessels on the entire South Branch. Despite lobbying and great protest, the U.S. Army Corps of Engineers, the U.S. Bureau of Public Roads, and the secretary of the army approved the use of a fixed bridge in March 1960. The bridge provided an uninterrupted throughway and set an important precedent for the Chicago River.

The Southwest Expressway Bridge opened to traffic in 1965. The steel and reinforced concrete bridge is 517 feet long and 174 feet wide and carries twelve lanes of traffic. There is an emergency 2'2" walk on each side. Ninety percent of the cost of this bridge was paid from federal gasoline-tax funds.

In August 1962, while still under construction, this bridge blocked a large commercial vessel delivering diesel fuel to the Pennsylvania Railroad terminal at 18th Street. After offloading locomotive fuel, the oil tanker *Polaris,* which had a capacity of thirty-five thousand barrels, had to be towed backward the way it had come through the South Branch to the Main Channel and then back out to Lake Michigan. Steering such a massive craft backward, even with the aid of tugboats, must have been quite a feat. Before the bridge, this ship could have continued down the South Branch through the Sanitary & Ship Canal to the Illinois River or, via the Calumet Sag Channel and Calumet River, could have returned to Lake Michigan.

| Dan Ryan (I-90/94) Expressway Bridge | | | | | |
|---|---|---|---|---|---|
| Opened | Bridge type | Designed by | Constructed by | Cost | Status |
| 1965 | Fixed, concrete and steel | Unknown | American Bridge Co., Division of U.S. Steel | $1,178,566 | Currently in use |

Looking east at the Dan Ryan Expressway Bridge in 2010. © Kevin Keeley

# SOUTH HALSTED STREET BRIDGES

LOCATION: 800 West, 2404 South; Halsted Street runs north and south, crossing the South Branch 4.4 miles from the river mouth at Lake Michigan.

HISTORICAL HIGHLIGHT: South Halsted was the site of the first modern vertical-lift bridge* in the world in 1894.

The current South Halsted Street Bridge was the first triple-pony-truss Chicago-type bascule. The new structure was developed as a solution to three limiting factors at this location. First, to the south there was an existing railroad viaduct crossing over the street that had to be taken into account. Second, the bridge approach could not exceed a slope of 4.65 percent for truck and streetcar traffic. Third, federal river-clearance regulations dictated a 16½-foot river clearance under at least 80 percent of the span. With only a few feet available within which to support the bridge roadway, City engineers developed a shallow-floor truss system. By incorporating a third truss in the center of the bridge, the entire superstructure could be composed of

---

*Although vertical-lift bridges had been invented previously, the third South Halsted Street Bridge was the first built on such a massive scale. J. A. L. Waddell developed the bridge and received Patent No. 506,571 on October 10, 1893.

small, light steel beams that could safely handle the bridge load requirements. As mentioned in the introduction, the new design extends only 3½ feet below the roadway, as compared to 5½ to 6 feet for a similar double-pony-truss bridge.

The bridge is 224 feet, 8 inches between the trunnions and offers a clear channel of 180 feet. It is 90 feet wide, with two roadways of 28½ feet each, and has 10½-foot-wide sidewalks on either side. Each of the two leaves weighs approximately 1,930 tons, which includes a counterweight of 1,300 tons. Interestingly, the concrete and cast-iron billets reused some of the counterweight materials of the old vertical-lift bridge. The bridge was of sufficient width to accommodate the later widening of South Halsted Street to 100 feet.

The first bridge built at South Halsted Street is reasonably represented by the image shown on the next page. It was taken from the contractor's letterhead and request for final payment from the City for completion of the bridge. The contract for the bridge determined that $8,000 would be paid to the contractors by residents and local business subscriptions, $3,000 would come from the City of Chicago, and the bridge and approaches would be completed by March 1, 1860. The contractors, Chapin & Howard, insisted on a $762 guarantee for subscribers on the list they considered doubtful and received personal notes from

| Current (fourth) South Halsted Street Bridge | | | | | |
| --- | --- | --- | --- | --- | --- |
| Opened | Bridge type | Designed by | Constructed by | Cost | Status |
| Sept. 13, 1934 | Chicago type, double-leaf bascule | Division of Bridges and Viaducts, City of Chicago | Mount Vernon Bridge Co. (superstructure), FitzSimons and Connell Co. (substructure) | $997,388 | Currently in use |

Looking east at the current South Halsted Street Bridge in 2007. © Laura Banick

Thomas O'Neil and James Iris for that amount as collateral. Work began on January 3, 1860, and soon the framework for the 121-foot bridge was near completion; on February 20, however, the commissioner of public works halted construction until further notice due to a court order. Apparently, two subscribers had sought recompense in the courts for extension of the street through their land west of the river.

The City argued that subscribing for the bridge meant that the two landowners knew full well that the street would be opened to the west by the usual means, that is, condemnation of property necessary to establish the right-of-way. According to the City, subscribing to the bridge precluded any claims for recompense for the land. The City began the proceedings for a special assessment and condemnation of the property to open

## First South Halsted Street Bridge

| Opened | Bridge type | Designed by | Constructed by | Cost | Status |
| --- | --- | --- | --- | --- | --- |
| July 1861 | Swing, wood, hand operated | Chapin & Howard | Chapin & Howard | $7,950 | Torn down in 1871 |

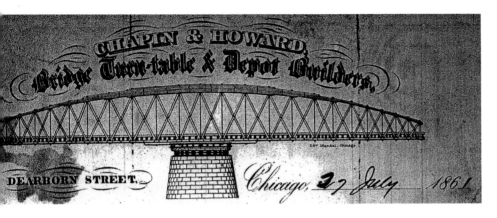

Chapin & Howard letterhead in 1861 with the representation
of the first South Halsted Street Bridge.

the street right-of-way. However, the process was delayed, as the City was forced into court on an appeal before the matter was resolved. Meanwhile, in March 1861, a year after the bridge was to have been completed, Chapin & Howard petitioned the Common Council to be allowed to complete the bridge, or at least receive payment for work done. The bridge work proceeded in June 1861 and was soon finished. In the meantime, the doubtful subscribers had indeed become insolvent or left the county. On July 27, 1861, Chapin & Howard sent a letter to collect from Thomas O'Neil and James Iris the final $762. This dragged out despite the City's having no legal involvement, and, in May 1862, O'Neil and Iris petitioned the Common Council for an investiga-

tion and relief of their guarantees. They argued that the halting of work by the commissioner of public works had added the delay that made collecting the guaranteed subscriptions impossible. They were denied relief on June 2 by the Common Council, and it is likely that this extended legal wrangling and petitioning of the City prompted the new policy, whereby the final subscription lists for a bridge would be turned over to the City to manage collection of funds and payment of contractors.

After ten years of use, this bridge "had become unsafe for travel," and it was removed in 1871.[1]

The second Halsted Street Bridge was built in 1872 and was a tubular iron-arch swing bridge. Built by the King Iron Bridge Company, the superstructure would have looked something like the drawing shown here from an 1866 King bridge patent. Zenas King, a former salesman of Thomas W. H. Moseley tubular iron bridges, began his own competing bridge company in 1858. He partnered with engineer Peter Frees, and they received a couple of patents to establish and create the very successful King Iron Bridge Company of Cleveland, Ohio.

A center pier and turntable supported the superstructure to create this 150-foot, double-arch swing bridge. The bridge had an 18-foot-wide roadway with two 6-foot-wide sidewalks.

The bridge soon suffered the ravages of Chicago River traffic. In 1882 it was badly damaged when the schooner *S. B. Wavel*

## Second South Halsted Street Bridge

| Opened | Bridge type | Designed by | Constructed by | Cost | Status |
|---|---|---|---|---|---|
| Early 1872 | Swing, iron, hand operated | King Iron Bridge Co. | King Iron Bridge Co. | $17,362 | Wrecked by steamer on June 30, 1892 |

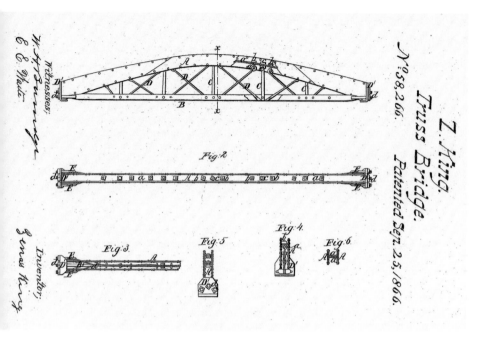

Zenas King truss bridge design, patented in 1866,
used on the second South Halstead Street Bridge.

struck the north end of the bridge with such force that it broke the locking bolts, swinging the bridge "completely around" and knocking it out of gear.[2] The impact also tore away several yards of railing and flooring. The City sought and received compensation from the vessel owners for the considerable amount of money needed to complete bridge repairs.

On June 31, 1892, the bridge met its demise in another collision. The eight-hundred-ton steamer *Tioga,* towed by the tug *Gardner,* had signaled for the bridge to open, and in response the bridge bell was rung. The ships proceeded. Bridge tender Thomas Curran later said he knew that the bridge was in need of repair—he had reported as much several days earlier—but did not dream it would stick as it did. Upon realizing that the bridge

was not going to open, the tug's driver pulled up alongside the steamer, stopped her engine, and reversed; however, the tug was unable to stop the heavy steamer's momentum in time. Both captains insisted that, as far as they were concerned, the accident was unavoidable. There was a terrific collision, and the ship knocked the bridge entirely off the turntable; it slammed over, falling diagonally; one end rested on the center pier, while the other fell onto the *Gardner.* The bridge was rent in two, a complete wreck. The pilothouse and smokestack of the tug were flattened, enveloping the whole scene in smoke and steam. Though several people were battered and bruised, amazingly, no one was killed.

After the wreckage was removed, subsequent plans to build another swing bridge were rejected by the War Department. Instead, a temporary pontoon footbridge was used for two years until a permanent solution could be devised. The temporary bridge was the scene of several incidents just a month later. A gang of ruffians opened the bridge each night to prevent pedestrians from crossing; then, using an old scow, they offered to ferry folks across for five cents per trip, claiming a canal boat had run into the bridge and it could not be swung. Those without coins or unwilling to pay were forced to walk to the 22nd Street Bridge. After several nights and a report in the newspaper, authorities put a stop to these shenanigans.

The third South Halsted Street Bridge followed, but City authorities were unable to gain federal approval for another swing bridge. Designs for the bridge caused significant debate involving City officials, the commissioner of public works, and the Sanitary District. It was finally proposed that the City build an unproven Waddell vertical-lift bridge. It would be the first large-scale vertical-lift bridge in the world and the City's second experiment in eliminating the objectionable swing-bridge design.

The *City of Paris* passing through the third South Halsted Street Bridge in 1900.

## Third South Halsted Street Bridge

| Opened | Bridge type | Designed by | Constructed by | Cost | Status |
|--------|-------------|-------------|----------------|------|--------|
| Mar. 22, 1894 | Waddell vertical lift, steel, steam, then electric powered | J. A. L. Waddell | Pittsburgh Bridge Co. | $242,880 | Removed by Strobel Steel Construction Co. in Dec. 1931 |

John Alexander Lowe Waddell's vertical-lift bridge had won a design contest held by the City of Duluth, Minnesota, in 1892. Though initially approved by the War Department, shipping interests ultimately prevented its construction in Duluth. Chicago took advantage of the award-winning design and built the world's first modern vertical-lift bridge at South Halsted Street.

The 160-ton thru-truss section spanned the river between two 220-foot steel towers. Steel cables attached to the bridge section ran to the tops of the towers over large sheaves (or pulleys) and connected with two 80-ton concrete counterweights suspended outside each tower. The bridge was balanced by the counterweights, which were operated with the aid of two small steam engines that raised and lowered the roadway between the two towers. The massive structure also offered a 155-foot clearance for river traffic. The bridge's high initial cost, frequent mechanical problems, cost of operation, and hulking structure, however, prevented the City from embracing the design. Chicago's experience with the bridge significantly injured the reputation of Waddell's design within the engineering community, and it would be more than a decade before another was built. Nonetheless, in 1908, Mayor Fred A. Bussel included this bridge as one of his "seven wonders of Chicago."[3]

This bridge effectively kidnapped five men on July 16–17, 1894. Four men, named Gutner, Ratcliffe, Brett, and Fox, were hurrying toward the bridge just as it was beginning to move upward. A police officer named Rosenkerns warned the men not to step onto the bridge, but they paid no attention and rode up with the draw. Once raised, the bridge refused to lower. The only means of escape was "a frail ladder" located on one of the towers, but none of the men, including similarly stranded bridge tender Patrick O'Keefe, was courageous enough to make use of

it.[4] The men were forced to make themselves as comfortable as possible. Charles Summers, an assistant engineer for the City, assisted with the repairs and ascended the rickety ladder *twice* during the ordeal to deliver the men food and water. A crowd gathered to view the spectacle, as workers labored throughout the night. Finally, after the men were held captive at an altitude of 160 feet for nearly thirty hours, a cogwheel in the operating machinery was repaired, and the men were released.

On April Fools' Day 1896, this bridge played quite a trick by dropping one of its four massive counterweights into the river. The bridge was being opened to let several boats pass; after it rose about 20 feet, a loud cracking was heard, and one of the counterweights near the top of the south tower came crashing down. It broke through the approach floor and fell into the river below. The suspected cause of the 150-foot free fall was either a loose bolt or the failure of an iron casting holding the eighty-ton weight. Scores of people were nearby, but, luckily, no one was injured. It took several days for a diver to hook the heavy weight so that it could be pulled from the river bottom. The retrieved counterweight was reattached, and the bridge returned to service.

Three years later, the bridge was twice disabled and closed for service. First, on October 2, 1899, a 1½-inch cable connected to the counterweights parted and took the bridge out of service for at least a day, at a repair cost of one thousand dollars. City engineer Ericson declared this bridge the most troublesome on the river and wished it would be replaced with a more "satisfactory type of bridge."[5] Just a month later, the bridge underwent repairs for another two days. The outcry from patrons was so great that William Loeffler, who lived in the neighborhood and was City clerk from 1897 to 1903 under Mayor Carter Harrison Jr., coordinated a solution. When he called the commissioner of public works, he was told that there was no money in the bridge fund and no way of repairing the bridge. Ericson gave the same answer. Loeffler went down to the bridge and talked with the residents and businessmen in the area. The consensus was that if the City would not provide a way across, they would do it themselves. With the help of local businessmen, a tug and sufficient fuel were procured for a temporary ferry. Loeffler returned to city hall and arm-twisted officials into supplying a crew so that the ferry was up and running by the next morning. The bridge was put back in service by the end of the following day.

In 1903 it was discovered that sand laid to crown the bridge deck under the paving blocks had added approximately 20 tons of load. Additional weight had not been added to the counterweights, and the unbalanced load added significantly to coal consumption and increased wear and tear on the bridge. When the bridge was later reconstructed without the sand, it operated on a steam pressure of 90 pounds from one engine, where before two engines at 110 pounds of pressure had difficulty lifting the deck. Removal of the sand also dramatically reduced the amount of repairs and interruptions.

In 1907 this bridge was converted to electric power by installing two 65-horsepower electric motors to replace the small steam engines. The City claimed a significant cost savings by eliminating the services of three engine operators and one coal shoveler. In addition, $170 per month was saved on coal, in favor of intermittent electrical usage amounting to roughly $50 per month, and only one bridge tender instead of two was required to operate the bridge. All together, the change to electric power saved the City $3,240 per year.

A year earlier, J. A. L. Waddell began his partnership with John Harrington, a skilled and innovative civil engineer, and the two redrew Waddell's vertical-lift design. The resulting Waddell-Harrington vertical-lift designs greatly improved on the bridge structure and operation. These became widely adopted by the railroads; seven such bridges in Chicago were built by the railroads, while an eighth at Torrence Avenue, built in 1938, was the only other vertical-lift bridge built by the City of Chicago.

The vertical-lift design had a potential problem with the changing cable weight, which shifted from the bridge side of the tower to the counterweight side and quickly took the bridge out of balance. As a clever solution, heavy chains were connected to the underside of each counterweight and looped up to attach to the bridge towers. As the counterweight is lowered, the additional weight of the steel cables drops over the bridge tower, adding to the counterweight. This is offset by a reduction in the chain weight that shifts to the bridge towers. This amazingly simple idea keeps the vertical-lift bridges in near-perfect balance.

In 1922 the end floor beams, sidewalk brackets, and vertical posts of the trusses received significant repairs under the direction of City engineer of bridge maintenance F. H. Avery, at a cost of $8,135. An additional $22,062 was spent to replace the counterweight cables. First replaced in 1912, the cables were in very good condition except for the ends connecting to the bridge and counterweights. At each connection, the cables passed through a long clamp and around a pin; upon disassembly, it was found that about half of the wire strands were broken and the ends badly corroded. All sixteen cables were changed (two at a time), halting river traffic. Two crews on twelve-hour shifts worked around the clock to complete the project over the next several weeks. The men labored through zero-degree temperatures and high winds, sheltered only by temporary structures built atop the towers. The first set of cables took ten days to replace, but with repetition the time improved so that replacing the last pair took just four days. In 1925 the original 65-horsepower motors were replaced with 75-horsepower motors, and the wood sheave houses were removed from the top of the bridge towers.

In a 1930 report to secure bonds to fund bridge improvements, City bridge engineers argued that this bridge should be removed because it posed the greatest interference to traffic of all City bridges. It was of inadequate strength to accommodate modern street traffic, since it was originally designed for the horse-and-buggy traffic of the 1890s. It had suffered the cumulative effects of insufficient bridge maintenance, and a detailed stress analysis found the bridge members to be very near their limit. The heavy streetcar and truck traffic exposed the bridge to stresses of up to forty-five thousand pounds per square inch on a daily basis. Simply repairing the bridge could not be justified, particularly since it offered a river channel of less than one hundred feet in width and caused an increased river current. The bridge's original roadway of thirty-four and a half feet was reduced in 1929 to minimize loads on the overstressed structure. At that time, the cantilevered sidewalks were replaced with new sidewalks inside the bridge truss that reduced the roadway, narrowing it to only two lanes. In December 1931, the South Halsted Street Bridge was closed to traffic and removed.

A temporary swing bridge built by John C. Paskins Company for $167,169.20 opened on December 20, 1931, to convey traffic while a Chicago-type bascule bridge was constructed. The temporary bridge was removed in 1934, after the new bridge was opened to traffic.

Balancing chains on a vertical-lift bridge. © Patrick McBriarty

# THROOP STREET BRIDGES

(No Current Bridge)

LOCATION: 1400 West, 2470 South. There is no longer a bridge at Throop Street (formerly Main Street), which runs north and south and has had two bridges crossing the South Branch 5.3 miles from the river mouth at Lake Michigan.

HISTORICAL HIGHLIGHT: Formerly Main Street, the road was renamed by grateful residents to be Throop Street, after Amos G. Throop, the Chicago lumberman and City treasurer, was instrumental in securing financing from New York to help rebuild the city after the Great Fire of 1871.

Looking east at the first Throop Street Bridge in 1900. Courtesy of MWRD.

The first bridge at Throop Street was a 152-foot swing bridge connecting Bridgeport with the southwest side of the city. Built by Fox & Howard, it was a twin-arch wood and iron–truss bridge that lasted an extraordinary thirty-five years. The bridge was replanked and repainted in 1878. In 1889 new turntable wheels were installed, and the bridge received a thorough reconstruction. In 1893, while the bridge at South Halsted Street was being reconstructed, traffic was deflected to Throop Street. In preparation for the additional traffic, the bridge approaches were rebuilt and replanked. The turntable was completely refurbished in 1897, receiving new tracks and rollers, and the protection pier was also repaired. In 1902 the Sanitary District removed the first Throop Street Bridge to make way for construction of a new bascule bridge.

Built by the Sanitary District, the second Throop Street Bridge was completed and turned over to the City on March 1, 1903. This Scherzer rolling-lift bridge saw a great deal of river traffic and, by the 1960s, was often closed for repair. Between 1963 and 1968, the bridge was closed for maintenance for a day or more at least a dozen times. This bridge opened more than

| First Throop Street Bridge | | | | | |
|---|---|---|---|---|---|
| Opened | Bridge type | Designed by | Constructed by | Cost | Status |
| 1868 | Swing, wood and iron, hand operated | Fox & Howard | Fox & Howard | $12,650 | Removed in 1902 |

## Second Throop Street Bridge

| Opened | Bridge type | Designed by | Constructed by | Cost | Status |
|---|---|---|---|---|---|
| Mar. 1, 1903 | Scherzer rolling lift, steel, electric powered | Scherzer Rolling Lift Bridge Co. | American Bridge Co. (superstructure), Lydon & Drews Co. (substructure) | $189,648 | Removed after new Loomis Street Bridge was built in 1978 |

three thousand times a year because of its low arch; much of its steelwork was below the 16½-foot river clearance at its center, forcing it to open for even river barges. It was one of the busiest bridges in the city. Removed in 1978, it was never replaced because a modern four-lane bascule bridge had been opened at Loomis Street one block west.

Looking east at the open second Throop Street Bridge in 1908.

# LOOMIS STREET BRIDGES

LOCATION: 1400 West, 2470 South; Loomis Avenue (formerly Deering Street) runs north and south, crossing the South Branch 5.3 miles from the river mouth at Lake Michigan.

HISTORICAL HIGHLIGHT: The current Loomis Street Bridge was the City's first attempt at using welded box girders in the construction of a bascule-bridge superstructure.

The first welded box-girder construction at Loomis Street ushered in the modern Chicago-type bascule era. Designed by City architect Jerome R. Butler Jr. and chief bridge engineer Henry Ecale, the bridge's "strikingly graceful and sleek visual image" was awarded the Most Beautiful Movable Bridge of 1978 award by the American Institute of Steel Construction.[1] The bridge became the prototype for the Columbus Drive Bridge built in 1982 and the Randolph Street Bridge built in 1984. Earlier techniques used rivets and later bolts to construct the bridge superstructure onsite, piece by piece. Improved welding technology allowed factory fabrication for final assembly at the job site, saving time and money.

This bridge was designed to carry area traffic after the Throop Street Bridge a block south was closed and removed. Soon thereafter, the Loomis Street Bridge opened. With the decline in manufacturing businesses in this industrial area, the Loomis Street Bridge has proved sufficient for the task of carrying local traffic across the South Branch.

Built in 1888, the first Loomis Street Bridge was two hundred feet long by thirty feet wide and of iron construction. It was hand operated and supported by a wood-pile center pier and piling-foundation approach piers.

This first Loomis Street Bridge nearly lost its bridge tender, Arthur Maynard, seven years after it opened. On June 12, 1895, Maynard saw what he thought was a runaway horse and buggy carrying an injured man. He ran into the street to attempt to stop it. After grasping the bridle of the horse, Maynard heard a bullet whistle over his shoulder. He quickly let go and turned to run, but the man in the buggy fired a second time and hit Maynard in the back just below the shoulder blade. Not losing "his nerve, however, even with a hole through him," Maynard managed to stagger to a patrol box and give the alarm before collapsing.[2]

The buggy sped from the bridge. Soon, Lieutenant Richard J. Moore of the Brighton Park Police Station was in pursuit. Moore had attempted to stop several suspicious-looking men several blocks before the bridge; they had turned out to be thieves. One of them brandished a revolver, and Moore shot him. Chasing on foot, Moore finally commandeered a buggy and pursued the thugs. Several miles beyond the Loomis Street Bridge, he lost them in the infamous Levy District, which was home

| Current (third) Loomis Street Bridge | | | | | |
|---|---|---|---|---|---|
| Opened | Bridge type | Designed by | Constructed by | Cost | Status |
| 1978 | Chicago type, double-leaf bascule | Division of Bridges and Viaducts, City of Chicago | Unknown | Unknown | Currently in use |

Looking northwest at the current Loomis Street Bridge in 2006. © Laura Banick

## First Loomis Street Bridge

| Opened | Bridge type | Designed by | Constructed by | Cost | Status |
|---|---|---|---|---|---|
| Apr. 30, 1888 | Swing, wood and iron, hand operated | Unknown | Shailer & Schniglau (superstructure), Chicago Dredge & Dock Co. (substructure) | $91,183 | Removed in 1904 |

Looking northeast at the first Loomis Street Bridge in 1902. Courtesy of MWRD.

to much of Chicago's vice and corruption. The three thieves were captured the next day and soon convicted of their crimes. Maynard was laid up for six weeks from his bullet wound before returning to work.

In 1904 the second Loomis Street Bridge, a Scherzer rolling-lift bridge built by the Sanitary District, replaced the first. The City assumed ownership upon completion and maintained and operated this bridge. The bridge was 178 feet long and 58 feet wide and was an electrically powered bascule with pile and timber approaches.

In 1940 the bridge received an entirely new road deck. By the 1960s, it was one of the most frequently operated bridges on the river, opening 3,624 times in 1968 alone. Although it had a 16½-foot clearance, the low angle of the support structure

## Second Loomis Street Bridge

| Opened | Bridge type | Designed by | Constructed by | Cost | Status |
|---|---|---|---|---|---|
| Oct. 17, 1904 | Scherzer rolling lift, steel, electric powered | Scherzer Rolling Lift Bridge Co. | Jackson & Corbett Co. (superstructure), Lydon & Drews Co. (substructure) | $231,249 | Removed in 1977 |

prevented barges from passing under it. In 1970 plans were submitted to replace the bridge because it was "antiquated and costly to keep maintained."[3] Delayed until 1977, the bridge was finally removed in favor of a modern bascule bridge.

Looking northeast at the open second Loomis Street Bridge in 1908.

# SOUTH ASHLAND AVENUE BRIDGES

LOCATION: 1600 West, 2601 South; Ashland Avenue runs north and south and crosses the West Arm of the South Branch 5.5 miles from the river mouth at Lake Michigan.

HISTORICAL HIGHLIGHT: About a half mile west of this South Ashland Avenue Bridge is the spot where the ailing Father Jacques Marquette spent the winter of 1674–75, before leaving that spring. Weeks later, Marquette would die on the eastern shore of Lake Michigan near Ludington, attempting to return to the French mission at St. Ignace, Michigan.

South Ashland Avenue has only one river crossing today, though it once bridged as many as four waterways. The street provides a study in Chicago River modifications and reengineering. The excerpt from a 1902 map of Chicago shows the street line of South Ashland Avenue, which crossed four waterways up until the mid-1930s. These were the West Fork of the South Branch, the Illinois & Michigan Canal, the culvert of the I&M Canal, and the South Fork of the South Branch.

Since this time, the I&M Canal, many of the river slips, and the canal along Pershing Road have been removed, and the South Fork up to 35th Street has been filled. After the opening of the Sanitary & Ship Canal to navigation in 1910, shipping on the I&M Canal declined rapidly, and it was officially closed in 1933. Soon afterward, most of the canal within the city limits was filled in. By the early 1940s, South Ashland Avenue had only one bridge crossing. The inventory of all the South Ashland Avenue bridges from north to south follows.

Map of the four South Ashland Avenue Bridge crossings in 1902 over the West Fork, I&M Canal, and I&M Culvert.

## The Crossing at the West Fork of the South Branch

This compact three-truss Chicago-type bascule bridge was built using the South Halsted Street Bridge design. The plans fitted the location and made immediate use of available Public Works Administration funds (which paid for the entire bridge) to avoid the delay of drawing new plans.

Just above the fork in the South Branch was the Leigh farm and homestead, originally called Lee's Place and also known as "Hardscrabble," which was settled in 1809. A Canal Commission map shows a bridge there as early as 1849, though there may have been two bridges at this location, one before and one after the flood of 1849. Because the area was outside the city limits at that time, there is very little information available on the first bridge here.

In 1871 a second South Ashland Avenue Bridge was built. It was removed in 1883 in favor of a third South Ashland Avenue Bridge with the cooperation of the City of Chicago and the I&M Canal Commission. This wood swing bridge was 160 feet long and 20½ feet wide, was hand operated, and had pile and timber approaches. To give some idea of the traffic in the area that year, a City survey conducted over a thirteen-hour period recorded 645 pedestrians and 287 vehicles crossing this bridge, making it the second least traveled of the twenty-seven City bridges surveyed

### Current (fifth) South Ashland Avenue Bridge over the South Branch

| Opened | Bridge type | Designed by | Constructed by | Cost | Status |
|---|---|---|---|---|---|
| Apr. 17, 1938 | Chicago type, double-leaf bascule | Division of Bridges and Viaducts, City of Chicago | Ketler-Elliott Co. (superstructure), FitzSimons and Connell Co. (substructure) | $1,163,607 | Currently in use |

### First, second, and third South Ashland Avenue Bridges over the South Branch

| Opened | Bridge type | Designed by | Constructed by | Cost | Status |
|---|---|---|---|---|---|
| Pre-1849 | Fixed, wood | Unknown | Unknown | Unknown | Unknown |
| 1871 | Swing, wood, hand operated | Fox & Howard | Fox & Howard | $5,000 | Removed in 1883 |
| 1883 | Swing, iron, hand operated | Detroit Bridge Co. | Detroit Bridge Co. (superstructure), Fox & Howard (substructure) | $18,318 | Removed in 1902 |

Looking northwest at the current (fifth) South Ashland Avenue Bridge in 2006. © Laura Banick

(the Madison Street Bridge was the busiest crossing, carrying more than fourteen times the number of vehicles and almost seventy times the number of pedestrians). In 1902 the Sanitary District removed this bridge in anticipation of a new bascule bridge.

The Sanitary District built this fourth bridge at South Ashland Avenue after widening the river as part of the construction of the Sanitary & Ship Canal. It was designed and patented by former Sanitary District engineer John Page.[1] The double-leaf, trunnion-bascule design used two cantilevered, counterweighted approaches, each hinging at the riverbank, and used a rack and pinion to raise each leaf (as shown in the patent drawings in the introduction). It was 264 feet long and 58½ feet wide. A key advantage of this design was that no part of the superstructure extended below the water level at any time during its use or operation. This avoided the larger foundation and more involved counterweight pits required by other bascule bridges; still, City engineers found it to be of poor construction and a less than robust design.

In 1904 new center locks were installed, the heel locks were repaired, and about one and a half tons of counterweights were

Looking west at the open fourth South Ashland Avenue Bridge in 1902. Courtesy of CDOT.

added to the north leaf. By 1906 the babbitt metal bearings at the four hinge points had been forced out of place by the great weight of the structure, requiring the installation of phosphor bushings. (Often used on bearings, babbitt metal was originally invented by Isaac Babbitt in 1839.) These repairs closed the bridge for almost a month, an action that could have been avoided had City of Chicago engineers been consulted during the bridge's design or construction. In 1925 this bridge prompted a rare commentary on the value of the Chicago-type bascule, expressed in the *Department of Public Works Annual Report:*

## Fourth South Ashland Avenue Bridge over the South Branch

| Opened | Bridge type | Designed by | Constructed by | Cost | Status |
|--------|-------------|-------------|----------------|------|--------|
| July 1, 1902 | Page bascule, steel, electric powered | John W. Page, civil engineer | Chicago Bridge & Iron Works (superstructure), Page & Schnable (substructure) | $160,106 | Removed Oct. 21– Nov. 11, 1936 |

*The most expensive repair ever made to a city bridge was carried out on the Ashland Avenue West Fork Bridge at a total cost of $155,107.00 [an amount greater than the initial cost of the bridge].*

*This included a modern type floor, complete new machinery and motors, new track plates and a temporary pontoon to take care of pedestrians during repairs.*

*This bridge has, however, furnished the best example of the fallacy of a city building bridges from competitive designs where the lowest first cost and lightest weight possible were the deciding factors of the design. As a result the City was saddled with structures doomed from the start to very expensive repairs and so light as soon to be overloaded by the greater weight and higher speed of modern traffic.*

*Experience in maintenance has shown that all of the patented type of bridges owned by the city due to their design, have made it necessary when extensive repairs were made, to place the span out of service and work overtime at an increase of about one-third of the labor cost in order to reduce the length of time the span was out of service.*

*The [Chicago-type] bridges designed by the City Bridge Division have avoided these troubles due to the fact that the experience gained during maintenance was applied to each successive new bridge and this experience is never available to the designers of patented bridges since they do not maintain them.[2]*

In 1936 the thirty-three-year-old Page bascule was removed and replaced by a Chicago-type bascule bridge.

## The Crossing at the Illinois & Michigan Canal

The first bridge over the locks at Bridgeport was constructed sometime before the flood of 1849 and was destroyed by that same disaster. A second bridge soon replaced it. Both of these bridges crossed the canal at a right angle and sat somewhat east of today's Ashland Avenue. In the 1880s, the canal commissioners and the City worked together to straighten the road below the West Fork of the South Branch, constructing a third bridge

### First, second, and third South Ashland Avenue Bridges over the Illinois & Michigan Canal

| Opened | Bridge type | Designed by | Constructed by | Cost | Status |
|---|---|---|---|---|---|
| Pre-1849 | Fixed, wood | Unknown | Unknown | Unknown | Destroyed by flood on Mar. 12, 1849 |
| 1849 | Fixed, wood | Unknown | Unknown | Unknown | Unknown |
| 1886 | Fixed, iron | Unknown | Pittsburgh Bridge Co. | $6,200 | Moved to temporary pier in Jan. 1909 |

4941 - 5-3-1914

Looking northeast at the fourth fixed bridge over the I&M Canal at South Ashland Avenue in 1914. Courtesy of MWRD.

## Fourth South Ashland Avenue Bridge over the Illinois & Michigan Canal

| Opened | Bridge type | Designed by | Constructed by | Cost | Status |
|---|---|---|---|---|---|
| Apr. 16, 1909 | Fixed, Pratt truss | Unknown | Milwaukee Bridge Co. (superstructure), John O'Brien (substructures) | $24,360 | Removed after 1933 |

in alignment with the street. Built in 1886, this fixed iron bridge spanned the I&M Canal until January 1909, at which point it was moved and used as a temporary pier to maintain traffic while a fourth bridge was constructed.

The fourth South Ashland Avenue Bridge over the I&M Canal was a steel span 127 feet long and 42 feet wide. It was used until the 1930s and the filling in of the I&M Canal. Afterward, the bridge was deemed to be no longer necessary and was removed shortly thereafter.

## The Crossing at the Culvert
## for the Illinois & Michigan Canal

The third crossing at South Ashland Avenue was over the culvert, or intake channel, of the I&M Canal, which fed the pumps at Bridgeport. Between 1880 and 1881, canal commissioners, with the cooperation of the City of Chicago, extended South Ashland

Avenue and constructed the first bridge across the culvert in line with Ashland Avenue. The Canal Commission built this pile-and-timber bent bridge, extending Ashland Avenue to the south. In 1908 the City replaced the bridge with a new bent bridge, paved with Shuman wood and asphalt pavement 3½ inches thick. About 1934 the I&M Canal channel was filled at this location, and the 78'-long × 48'-wide fixed wood bridge was removed.

## The Crossing at the South Fork
## of the South Branch

The original South Fork of the South Branch meandered west just below Pershing Road (known then as 39th Street) and was not much more than a creek; rather, it was part of the swampy landscape that would become the future site of the Union Stock Yards. In the 1860s, the stream was dredged and widened for vessels. The first bridge at this location was a pontoon pivot pier

### South Ashland Avenue Bridges over the culvert of the Illinois & Michigan Canal

| Opened | Bridge type | Designed by | Constructed by | Cost | Status |
|---|---|---|---|---|---|
| 1880 or 1881 | Fixed, wood | Unknown | Canal Commission | Unknown | Removed in Nov. 1908 |
| Dec. 28, 1908 | Fixed, wood | Unknown | FitzSimons and Connell Co. | $5,293 | Removed in the mid-1930s |

### First South Ashland Avenue Bridge over the South Fork of the South Branch

| Opened | Bridge type | Designed by | Constructed by | Cost | Status |
|---|---|---|---|---|---|
| 1871 | Swing, wood, hand operated | Fox & Howard | Fox & Howard | $5,000 | Removed on Dec. 16, 1907 |

Looking west at the second South Ashland Avenue Bridge over the South Fork in 1917. Courtesy of MWRD.

## Second South Ashland Avenue Bridge over the South Fork of the South Branch

| Opened | Bridge type | Designed by | Constructed by | Cost | Status |
|---|---|---|---|---|---|
| Oct. 28, 1908 | Swing, steel, electric powered | Unknown | King Bridge Co. (superstructure), FitzSimons and Connell Co. (substructure) | $85,357 | Bubbly Creek filled to east of Ashland Avenue and bridge removed in 1920 |

moved from 22nd Street. It was 152 feet long by 18½ feet wide and was installed as a temporary bridge by FitzSimons & Connell Company at a cost of $9,991. It was used until 1907, when it was removed to make way for construction of a new swing bridge.

The new bridge was completed in 1908. This steel swing bridge was 176 feet long and 42 feet wide and was powered by electricity. In 1913 the stream was closed to navigation, the bridge tenders were transferred, and the span was closed for repairs. As shown here in the 1917 image, a cofferdam sits in front of the bridge, while behind it the stream is being filled in. The bridge was removed in 1920 following a City ordinance that permitted the filling of Bubbly Creek east to Ashland Avenue. Traffic was temporally diverted to a plank roadway, while streetcars continued using the tracks without diversion. The roadway was paved with granite blocks and reopened. By the early 1930s, the South Fork of the South Branch was further filled north just below 35th Street, which is how it remains today.

# THE BRIDGES OF THE

## (FROM SOUTH TO NORTH)

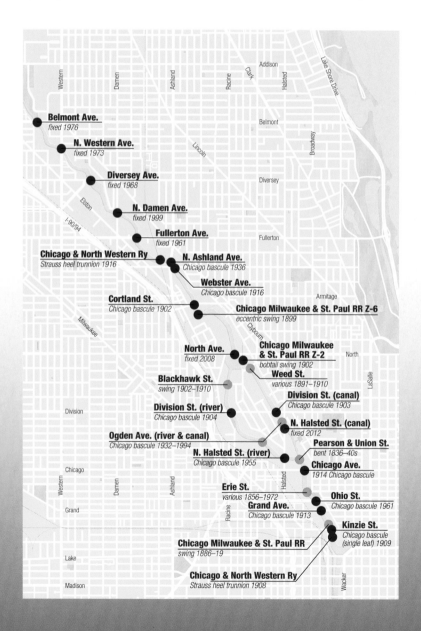

**Belmont Ave.**
*fixed 1976*

**N. Western Ave.**
*fixed 1973*

**Diversey Ave.**
*fixed 1968*

**N. Damen Ave.**
*fixed 1999*

**Fullerton Ave.**
*fixed 1961*

**Chicago & North Western Ry**
*Strauss heel trunnion 1916*

**N. Ashland Ave.**
*Chicago bascule 1936*

**Webster Ave.**
*Chicago bascule 1916*

**Cortland St.**
*Chicago bascule 1902*

**Chicago Milwaukee & St. Paul RR Z-6**
*eccentric swing 1899*

**North Ave.**
*fixed 2008*

**Chicago Milwaukee & St. Paul RR Z-2**
*bobtail swing 1902*

**Weed St.**
*various 1891–1910*

**Blackhawk St.**
*swing 1902–1910*

**Division St. (canal)**
*Chicago bascule 1903*

**Division St. (river)**
*Chicago bascule 1904*

**N. Halsted St. (canal)**
*fixed 2012*

**Ogden Ave. (river & canal)**
*Chicago bascule 1932–1994*

**Pearson & Union St.**
*bent 1836–40s*

**N. Halsted St. (river)**
*Chicago bascule 1955*

**Chicago Ave.**
*1914 Chicago bascule*

**Erie St.**
*various 1856–1972*

**Ohio St.**
*Chicago bascule 1961*

**Grand Ave.**
*Chicago bascule 1913*

**Kinzie St.**
*Chicago bascule (single leaf) 1909*

**Chicago Milwaukee & St. Paul RR**
*swing 1886–19*

**Chicago & North Western Ry**
*Strauss heel trunnion 1908*

North Branch map. Courtesy of Chicago CartoGraphics.

The North Branch of the Chicago River stretches fifteen miles within the Chicago city limits. The five-and-a-half-mile portion from the forks up to Belmont Avenue is considered navigable and hosts seventeen bridges, which have been subject to federal oversight since 1890. The initial development of this portion of the city occurred slightly later than much of the rest of Chicago and tended toward residential, away from the river. Similar to the other branches, the North Branch was repeatedly widened, deepened, and straightened for shipping and industry.

The most significant alteration was the addition of the North Branch Canal that created Goose Island, the only island in the Chicago River. Originally referred to as the Ogden Canal, the waterway cut between Chicago Avenue and North Avenue. The area was purchased from the I&M Canal Commission by a group of private investors led by William B. Ogden. The canal was completed in the 1850s as part of a real-estate development

# NORTH BRANCH

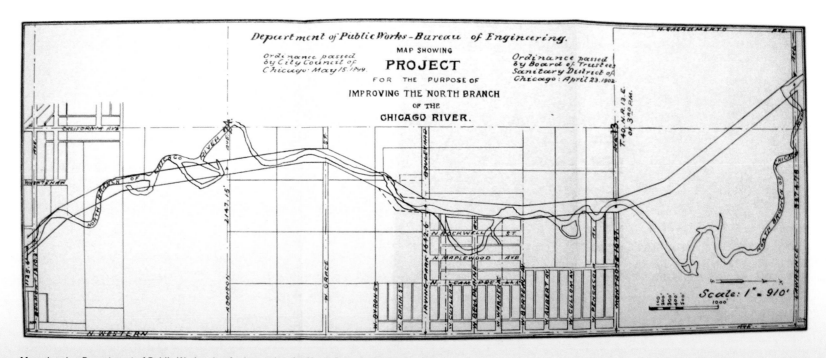

Map showing Department of Public Works plan for improving the North Branch channel in 1903.

scheme to create additional waterfront property. The land was then subdivided into business and residential parcels and sold.

By the 1880s, the North Branch had become home to multiple coal yards, tanneries, distilleries, iron- and brickmaking firms, and at least two shipyards. Today, this area, which includes Goose Island, is often referred to as the North Branch Industrial Corridor and still retains a commercial and industrial base. As an auxiliary channel, the North Branch never carried the same volume of shipping as the Main Channel and South Branch waterways. During the twentieth century, the North Branch experienced the most dramatic decline in commercial shipping.

In 1910 the Sanitary District added the eight-mile North Shore Sanitary Canal. It originates at the lakefront in Wilmette and runs south, connecting with the North Branch near Foster Avenue. A dam and sluice gate at the northern end just west of Sheridan Road control the flow of water from Lake Michigan into the North Branch. Though the canal is considered nonnavigable, small recreational boats are known to use it.

The replacement of drawbridges with fixed spans on the navigable portions of the North Branch was a long-term goal of the City of Chicago. Following the 1960s, however, a significant drop in drawbridge use and the long-sought after federal approval finally supported its implementation, thereby reducing bridge-maintenance and operational costs. Today, all other City drawbridges over the North Branch and North Branch Canal have been permanently fixed in place and no longer operate or have been replaced by newer fixed bridges. Like the rest of the Chicago River, gentrification has encroached on the area, and much fewer manufacturing companies remain. The shift to primarily retail and residential land use in the North Branch occurred rapidly over the past three decades.

# KINZIE STREET BRIDGES

LOCATION: 423 West, 400 North; Kinzie Street runs east and west, crossing the North Branch 1.8 miles from the river mouth at Lake Michigan.

HISTORICAL HIGHLIGHT: Kinzie Street was the location of Chicago's very first bridge (built in 1832), a river fire (in 1899), and a flood that shut down the city's center for more than a week (in 1992).

This sixth Kinzie Street Bridge was built in 1909 and was one of two single-leaf first-generation Chicago-type bascule bridges. The finishing elements of the bridge were added the following year: a danger-signal system was installed at the west approach, and the bridge deck was paved.

Further improvements and repairs began in 1911 when the automatic gate was repaired, the pavement on the approaches patched, and the bridge houses painted. The Chicago, Milwaukee & St. Paul Railway tracks at the foot of the west approach were raised about nine inches. This reduced the approach grade to the bridge from 6 percent to about 4 percent. The new grade was paved in granite blocks by the Bureau of Streets and charged against the bridge-repair fund that same year. In 1931 the eastern bridge house was elevated above the sidewalk and rebuilt to greatly improve visibility for bridge operations. The new frame house, with an asbestos shingle roof and sides, was placed on additional steel bracing and received new electrical and mechanical equipment and a new Arcola heating system. In 1939 the upper two stories of the State Street bridge house were moved to the southwest corner of the Kinzie Street Bridge and installed on a new concrete foundation with a new heating system. The dilapidated watchman's house at the west approach was replaced. In addition, renewal of the floor system, bridge deck, and sidewalks, begun the year before, was completed. In 1953 the Kinzie Street Bridge was reconfigured for one-person operation, which resulted in the elimination of four bridge-tender positions, a savings to the City of sixteen thousand dollars per year. In 1969 the bridge was redecked and repaired by City workers.

Its low river clearance made this bridge one of the City's most active bridges, opening an average of five thousand times per year until 1998. In November of that year, the street was closed while the entire structure was raised five feet to provide a new river clearance of twenty feet. After reopening it required only one-twentieth of the former number of openings. This was the last City bridge on the Chicago River to be manned twen-

| **Current (sixth) Kinzie Street Bridge** | | | | | |
|---|---|---|---|---|---|
| Opened | Bridge type | Designed by | Constructed by | Cost | Status |
| May 10, 1909 | Chicago type, single-leaf bascule, steel, electric powered | Alexander von Babo, city engineer of bridge design | John J. Gallery (superstructure), Great Lakes Dredge & Dock Co. (substructure) | $199,750 | Currently in use |

Looking south at the current Kinzie Street Bridge in 2007, with the open North Western Railroad Bridge in the background. © Laura Banick

ty-four hours a day, seven days a week. Now this and the other North Branch drawbridges have been fixed in place and are no longer operational.

This Kinzie Street Bridge has been the scene of several major incidents in recent history. For instance, in 1979 Leon Sutton got into trouble after disregarding the bridge's warning lights. He sped west across the bridge but got trapped on the far side by the traffic gate. Meanwhile, bridge tender Sam Vinci, apparently not having seen Sutton's car, began opening the bridge for a river barge. Sutton's car rested half on and half off the bridge. As the bridge opened and the back end of the car was lifted into the air, Sutton realized his peril and jumped to safety just moments before his car dropped off the end of the bridge into the river. Both driver and bridge tender were cited in the incident. Sutton was ticketed for running a red light, while Vinci was suspended for a week for failure to follow bridge-lift operating procedures.

In 1987 a similar occurrence—with much more dire consequences—occurred on a Tuesday evening when a westbound taxi got trapped inside the warning gate. The cab sat with its back wheels on the bridge, unnoticed by the bridge tender. As the bridge was raised, passenger Reva Hawk, thirty-one, jumped from the cab, somehow landing on the street instead of in the river. The bridge flipped the cab upside down, leaving the vehicle resting on the west approach with the front of the cab dangling over the river. Hawk quickly pulled her friend Jane Williams, twenty-four, from the back of the cab. Williams had sustained a broken arm and bruises from being thrown around inside the cab, but escaped with her life. The raised bridge obscured the view of the overturned cab from the bridge tender, and once the ship traffic cleared, the bridge was lowered. Hawk and a male bystander began screaming at the bridge tender to stop lowering the bridge, but to no avail; the cab was crushed, killing taxi driver Ju H. Bang. When tested almost four hours after the incident, the bridge tender still had a blood alcohol reading of 0.152, which led to his firing by the City. Criminal charges were not filed in the accident because there is no law against operating a drawbridge under the influence.

On April 13, 1992, the river at Kinzie Street saw frantic City workers throwing mattresses and sacks of concrete into a swirling vortex in the river. New protection pilings for the Kinzie Street Bridge driven seven months before had ruptured a century-old freight tunnel twelve feet below the riverbed. The piling work was part of a City contract with Great Lakes Dredge & Dock Company to replace piling clusters at a half-dozen bridges. City employees on an inspection tour had failed to examine the Kinzie Street work earlier in the year due to an inability to find parking. Back in January 1992, cable television workers on a videotape survey discovered three wood pilings driven through one side of the tunnel. Though there was no water, a drift of knee-high silt and river mud had oozed into the tunnel. Their call attempting to inform the City failed due to a municipal reorganization that changed departmental numbers and further delayed informing the proper authorities.

That March a City engineer personally viewed and photographed the ruptured tunnel. The film was dropped at a local Walgreens. The prints of the damage were not picked up until a week later, at which time they were presented to several mid-level City officials. In early April, a repair estimate of seventy-five thousand dollars was considered too costly, so arrangements for additional quotes were made, and several contractors were

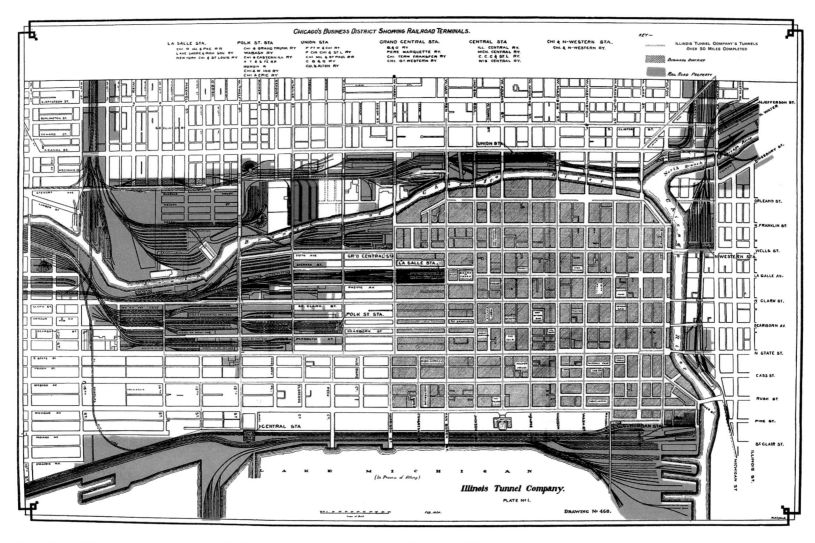

Map of the freight tunnel system (in red) beneath Chicago's downtown and railroad lines (in green) in 1904.

scheduled to visit the tunnel on Monday, April 14. Officials did not feel any sense of urgency, given that the tunnel had held for the past six months.

However, on Sunday, April 13, the tunnel gave way, and soon the sixty-two-mile network of tunnels below Chicago's downtown began filling with river water. By Monday morning, the water in the tunnels began to seep into the multilevel basements of major downtown buildings around the Loop. As a safety measure, electric utility company Commonwealth Edison (better known as ComEd) shut down large portions of the power grid, as the water threatened electrical and phone distribution systems in building basements. The resulting power outages forced the evacuation of much of the city's center, effectively shuttering downtown buildings and businesses for more than a week. After a couple of days, the hole in the tunnel was plugged, and a week later basements were pumped out and power restored. The incident, caused by a series of bureaucratic miscues, was estimated to have caused eight hundred million dollars in damage and resulted in the firing of at least two City engineers.

More recently, the Kinzie Street Bridge was the scene of a rather messy incident when, on August 8, 2004, the tour boat *Little Lady* was deluged with septic waste from a tour bus while passing under the bridge. The bus driver allegedly emptied the bus's holding tank (an estimated eight hundred pounds of human waste) while driving over the bridge. Much of the contents descended through the open steel grate of the bridge deck, into the river, and onto the upper deck of the *Little Lady*, which was carrying 117 passengers. Chicago Police determined that the bus had been chartered by the Dave Matthews Band. Although it was reported no member of the band was on the bus, the group made unsolicited donations of fifty thousand dollars to the Friends of the Chicago River and fifty thousand dollars to the Chicago Park District. Five months later, the bus driver turned himself in and pleaded guilty to reckless conduct and unlawful discharge of contaminated waste. He received eighteen months

of probation and a fine of ten thousand dollars and was required to perform 150 hours of community service. In April 2005, the Dave Matthews Band paid two hundred thousand dollars to settle a civil lawsuit stemming from the incident.

In 1832 Samuel Miller built Chicago's very first bridge. A simple wood footbridge intended to connect patrons with his tavern, it quickly became a modern convenience for all. This river crossing was motivated by self-interest: Miller wanted to gain business and generate profit competing against Wentworth's Tavern on Wolf's Point on the west side of the river. By 1833 this bridge was improved to accommodate horse and wagon teams. A lack of funds allowed this 120'-long × 10'-wide bridge to fall into disrepair until an appropriation funded its repair in 1837. In 1839 it was removed and replaced with a moveable bridge.

The second Kinzie Street Bridge was a pontoon swing bridge. The approaches on either bank were supported by pilings driven into the riverbed. Two semifloating aprons connected the approaches with a floating draw in the center. Each apron was supported by pilings at one end and a wood float on the other. Between these aprons at the bridge's center was the floating pier, or *draw*, supported by wood pontoon floats. Releasing pins on three corners allowed bridge tenders to pivot the draw on the fourth corner with the use of chains attached to the free end of the bridge. The chains were of sufficient length to reach either shore, drop to the river bottom, and not snare passing ships. As one

| First Kinzie Street Bridge | | | | | |
|---|---|---|---|---|---|
| Opened | Bridge type | Designed by | Constructed by | Cost | Status |
| 1832 | Fixed, wood bent | Samuel Miller | Samuel Miller | Unknown | Removed in 1839 |

## Second Kinzie Street Bridge

| Opened | Bridge type | Designed by | Constructed by | Cost | Status |
|--------|-------------|-------------|----------------|------|--------|
| 1839 | Pontoon swing, wood, hand operated | Unknown | L. Price and R. Freeman | Unknown | Destroyed by flood on Mar. 12, 1849 |

could imagine, tending these bridges could be rather strenuous without adding in the dirty job of handling lengths of chain that had been lying on the muddy river bottom. Opening and closing the bridge usually took between twenty minutes and a half hour.

On March 12, 1849, a flood destroyed this and all other Chicago bridges, as recounted in "The Flood of 1849" in the introduction.

Rebuilding after the flood, a third, more substantial, Kinzie Street Bridge was introduced. It reused the three floats from the old bridge and cost the City $917.55; the remaining $358.45 that funded the bridge was paid for by local subscriptions. With piling approaches on either side, the draw was supported by floats at one end and a light turntable on a circular pier of pilings at the other. A single semifloating apron connected with the float end of the draw. The bridge elevated the roadway about five feet above the water, nearly equal to the grade of the street. This left about a four-foot clearance under which small boats could pass.

By 1853 the old floats were worn out, and residents petitioned the Common Council, relating that pumping "day and night" by the bridge tender could not prevent the floats from frequently sinking.[1] They characterized the bridge as "dilapidated" and "at present entirely useless" and recommended that a new bridge be built.[2] Instead, it was repaired. In 1857 it was lengthened and refurbished to match a recent widening of the river for ship traffic. Finally, in 1859, this bridge was removed and replaced.

The fourth Kinzie Street Bridge was a swing bridge and a vast improvement over the turntable pontoon bridges. Its more substantial structure carried the dramatically increasing traffic and commerce of the city. The bridge was operational for eleven years, after which it was removed and replaced by a new swing bridge.

The fifth Kinzie Street Bridge, built in 1870, was a Howe-truss, hand-operated swing bridge, 170 feet long by 31½ feet wide.

## Third Kinzie Street Bridge

| Opened | Bridge type | Designed by | Constructed by | Cost | Status |
|--------|-------------|-------------|----------------|------|--------|
| Sept. 1849 | Pontoon turntable swing, wood, hand operated | Derastus Harper, city superintendent of public works | Derastus Harper, city superintendent of public works | $1,276 | Removed in 1859 |

## Fourth Kinzie Street Bridge

| Opened | Bridge type | Designed by | Constructed by | Cost | Status |
|---|---|---|---|---|---|
| Nov. 3, 1859 | Swing, wood, hand operated | Newton Chapin & Co. | Newton Chapin & Co. | Est. $15,000 | Removed in 1870 |

Looking north from Randolph Street at the west approach of the Lake Street Bridge and the Kinzie Street Bridge in the background, pre-1870. Courtesy of the Chicago Public Library, Special Collections and Preservation Division.

On April 17, 1899, the Chicago River at Kinzie Street mysteriously burst into flames and burned fiercely for more than fifteen minutes. The flames rose to a height of fifty feet and set fire to both the Kinzie Street Bridge and the Chicago & North Western Railway Bridge just to the south. The cause of the ignition was undetermined, but the Kinzie Street bridge tender smelled gas just before the fire started, and the firefighters concurred, adding that "the river's surface was covered with oil."[3] It was suspected that the oil came from an open sewer pipe that emptied into the river under the bridge. A Chicago & North Western train crossed the bridge just prior to the start of the fire, and it was believed that a spark from the coal-fired steam engine may have triggered the blaze. Both bridges were saved, and, as reported in the *Chicago Tribune*, "the damage wrought by the fire" was as follows:

| | |
|---|---|
| To Kinzie street bridge and approaches | $5,000 |
| To the Northwestern railroad bridge | 1,000 |
| To John Drucker's warehouse | 200 |
| To the docks | 200 |
| To the river | Nominal[4] |

## Fifth Kinzie Street Bridge

| Opened | Bridge type | Designed by | Constructed by | Cost | Status |
|---|---|---|---|---|---|
| 1870 | Swing, wood and iron, hand operated | Fox & Howard | Fox & Howard | $15,850 | Removed on Dec. 16, 1907 |

The open Chicago & North Western Railway, fifth Kinzie Street, and Chicago Milwaukee & St. Paul Railroad bridges from south to north. Courtesy of the Chicago Maritime Museum.

The bridge house, western approach, roadway, and sidewalks of the Kinzie Street Bridge were completely rebuilt after the fire.

Four years later, the planking on this bridge had to be repeatedly repaired due to the heavy wagon-team traffic crossing the bridge. The old timbers in the center pier gave way beneath the center step, and the bridge was taken out of service on September 11. The span was jacked up for repair; the center pier was practically rebuilt from the waterline up, new bridge seats were installed, and fifty-three piles were added to the pier protection and bridge approaches. In 1905 the *Public Works Annual*

*Report* stated that "the traffic on this bridge is probably heavier than any other bridge in Chicago." The bridge was again taken out of service, the deck was removed, and the bottom chords were straightened.[5] A system of top and bottom laterals was installed, and six tons of steel-plate reinforcements were riveted to the chords. The truss rods, main braces, and counterbraces were replaced, and the old approaches and bridge seats were removed and completely rebuilt. In addition, the sewer under the west approach was rebuilt, and the bridge was redecked and given two coats of paint before it was put back into service.

On December 11, 1907, the temporary pontoon bridge used at North Avenue was opened at Kinzie Street. The next day, traffic was halted on the old swing bridge, and the Great Lakes Dredge & Dock Company began removal. Construction of a new bascule bridge followed. On January 30, 1909, the pontoon of the temporary bridge sank to the bottom of the river while the new bridge was still under construction. The pontoon bridge was refloated, the motors and electrical systems were replaced, and the bridge was put back into service on February 4. This temporary pontoon bridge was removed from use and stored in the turning basin by North Avenue after completion of the new Kinzie Street Bridge.

# GRAND AVENUE BRIDGES

LOCATION: 462 West, 450 North; Grand Avenue (formerly Indiana Avenue) runs east and west, crossing the North Branch two miles from the river mouth at Lake Michigan.

The current Grand Avenue Bridge is a second-generation Chicago-type bascule bridge. Not long after this double-leaf bascule was built, the nineteen-foot-wide bridge was considered too narrow and the cause of significant automobile congestion; however, proposals to widen or replace the bridge in the 1920s, 1940s, and 1960s never got past the planning stage. Construction of the expressways, particularly the opening of the Ohio Street Bridge and feeder ramp in 1961, shifted traffic away from Grand Avenue. As evidence, repair work to reconstruct an outdated approach and viaduct over the railroad tracks (portions of which dated back to the 1870s) that left the bridge closed to traffic from August 27, 1975, to September 19, 1977, resulted in little to no complaint. This second Grand Avenue Bridge has recently passed its hundredth birthday and is still in use today. In the early 2000s, the two bridge leaves were fixed with an I-beam locking them together at the center, and maintenance of the operating equipment was abandoned.

The first bridge at Grand Avenue over the North Branch was a common Howe-truss bridge. The company of Fox & Howard built this 163'-long × 32'-wide wood and iron swing bridge. Fox & Howard were master builders of the commonly used combination wood and iron superstructures shown here. This bridge was built with the assistance of the Chicago & North Western Railway Company, which constructed a viaduct west of the river to carry the road over its tracks. The stone-pier and iron-truss bridges of the viaduct extended from the west abutment of the bridge to Des Plaines Street. In 1895 this bridge received a reinforced floor system that carried electric streetcar traffic, just as had happened at Erie Street.

Combination wood and iron swing bridges like this one typically lasted only about twenty-five years; this bridge, however, survived for forty. In 1907 the dilapidated condition of the trusses demanded that the bridge be swung over the protection pier so it could be repaired. First, forty-one pilings were driven under the bridge chords, providing the necessary support for the bridge to be jacked up. Then, broken and decayed truss members were replaced, several vertical rods were added to each truss, the structure was realigned and reinforced, and the bridge was given one complete coat of paint and put back into service. On June 3, 1910, the FitzSimons & Connell Company began dismantling this bridge. The center-pier protection and piling clumps were removed, and the river was dredged to 22 feet. The work was completed on July 21, 1910, at a total cost of $4,857.20.

| Current (second) Grand Avenue Bridge | | | | | |
|---|---|---|---|---|---|
| Opened | Bridge type | Designed by | Constructed by | Cost | Status |
| Dec. 13, 1913 | Chicago type, double-leaf bascule, steel, electric powered | Division of Bridges and Viaducts, City of Chicago | Strobel Steel Construction Co. (superstructure), FitzSimons & Connell Co. (substructure) | $195,141 | Currently in use |

Looking southeast under the Ohio Street Bridge at the current Grand Avenue Bridge in 2007. © Laura Banick

## First Grand Avenue Bridge

| Opened | Bridge type | Designed by | Constructed by | Cost | Status |
|---|---|---|---|---|---|
| 1869 | Swing, wood and iron, hand operated | Fox & Howard | Fox & Howard | $48,800 | Removed between June 3 and July 21, 1910 |

Looking northwest at the first Grand Avenue Bridge in 1909. Courtesy of the Chicago History Museum.

# OHIO STREET BRIDGE

LOCATION: 500 West, 560 North; Ohio Street runs east and west, crossing the North Branch 2.1 miles from the river mouth at Lake Michigan.

The Ohio Street Bridge serves a major artery connecting the nearby North Side to the Kennedy Expressway. The Kennedy, constructed in the late 1950s, opened to traffic on November 5, 1960. Initially named the Northwest Expressway for the direction in which it moved away from the city, the Chicago City Council voted unanimously on November 29, 1963, to rename the expressway after the recently assassinated President John F. Kennedy.

This river crossing comprises twin side-by-side double-leaf bascule bridges, owned and maintained by the State of Illinois and operated by the City of Chicago. It is the first bridge built at this location. The Ohio Street Bridge ties into the expressway to greatly relieve traffic on the North Side. As with the Congress Street Bridge, the two halves of this bridge are staggered to reduce the overall length of the structure. Each is a three-lane bridge, one carrying eastbound and the other westbound traffic. Both bridges have bridge houses strategically placed on opposite corners of the roadway.

| Ohio Street Bridge | | | | | |
|---|---|---|---|---|---|
| Opened | Bridge type | Designed by | Constructed by | Cost | Status |
| 1961 | Twin Chicago type, double-leaf bascules, steel, electric powered | Division of Bridges and Viaducts, City of Chicago | Strobel Steel Construction Co. (superstructure), FitzSimons & Connell Co. (substructure) | $195,141 | Currently in use |

Looking northwest at the Ohio Street Bridge in 2008. © Laura Banick

# ERIE STREET BRIDGES

(No Current Bridge)

LOCATION: 610 West, 670 North. There is no bridge today at Erie Street, which runs east and west. Between 1860 and 1971, this location had three bridges, crossing the North Branch 1.9 miles from the river mouth.

This first bridge at Erie Street was a pontoon swing bridge, which was one-quarter the cost of a regular swing bridge. At the time, pontoon swing bridges, although still common and reliable, were being replaced by the newer swing-bridge designs. Locations farther away from the city's center, including, at that time, Erie Street, typically received the most economical bridges as their first bridges. In 1865 a spring flood swept the bridge away, washing it down the river to Grand Avenue. It was salvaged, rebuilt, and served for another six years at Erie Street before finally being replaced in 1871.

In April 1871, Erie Street received its second bridge, which was two hundred feet long by thirty-two feet wide. This bridge highlighted the difficulty and significant land requirements that go along with the swing design, especially in an area with high property values. The City leased the right to swing the bridge over a strip of private property at the southwest corner of the riverbank for $37.50 a month. By early 1880, the lease had expired, and no additional payments from the City were forthcoming. In October

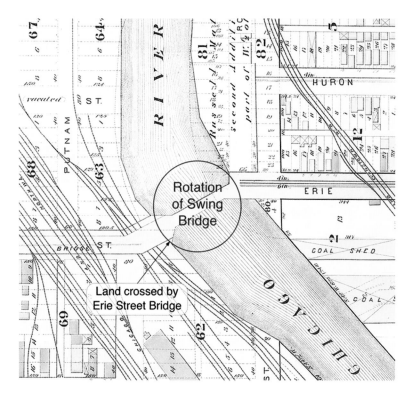

Map of the second Erie Street Bridge in 1886.

the property owner, in frustration, planted posts to prevent the swinging of the bridge over his land and petitioned the City to pay rent at the old rates or purchase the property for $15,000. Ships at a private dock north of the bridge regularly obstructed the bridge from swinging in the other direction, creating an impasse that remained unresolved well into 1881. It was once reported

| First Erie Street Bridge | | | | | |
|---|---|---|---|---|---|
| Opened | Bridge type | Designed by | Constructed by | Cost | Status |
| 1856 | Pontoon swing, wood, hand operated | Unknown | Unknown | $5,000 | Removed in 1871 |

## Second Erie Street Bridge

| Opened | Bridge type | Designed by | Constructed by | Cost | Status |
|--------|-------------|-------------|----------------|------|--------|
| April 1, 1871 | Swing, wood, hand operated | Fox & Howard | Fox & Howard | $30,000 | Collapsed under its own weight on May 18, 1908 |

that the bridge had been left open for more than fourteen days; citizens agitated to take up a subscription to pay the landowner, or reasoned that, if a bridge was not required at Erie Street, it should be taken away to prevent people from going out of their way for nothing. The City purchased the land soon after and resolved the conflict.

In 1888 the turntable was rebuilt, and the bridge house was moved to a more appropriate location on the other side of the bridge. The entire structure was repainted, and new pilings were driven to revitalize the entire south half of the protection pier. In 1895 the bridge floor was reinforced so that it could handle electric-streetcar traffic. In 1899, after the steamer *Charles Eddy* hit the bridge, knocking it south sixteen inches and bending the new flooring girders about nine inches, the City made temporary repairs. The following year, the Army Corps of Engineers widened the east draw by removing the projecting angle of the shore and dredging the channel. Afterward, the City entirely rebuilt the eastern approach of the bridge.

By May 18, 1908, the thirty-seven-year-old bridge, which had been "regarded as weak for some time," came to a safe end thanks to the "exceptional judgment" of Erie Street bridge tender John Meyers.[1] A few days earlier, a main support timber had rotted in two, and the bridge was thrown out of alignment.

Around five in the afternoon, after a heavily laden coal wagon crossed the bridge, a cracking sound was heard, and new breaks in the floor were observed. Fearing the structure might fall into the river, Meyers attempted to turn it, but he soon discovered something wrong—rotating the key had no effect. He hailed a Great Lakes tug to assist by pulling the bridge open. The tug attached a rope and began pulling, and as the bridge came off the abutments, the volume of the cracking noises increased. No sooner had the tug pulled the sagging bridge over the protection pier than the main timbers supporting the bridge split down the middle and shot up through the bridge floor. The timbers connecting the top and bottom trusses snapped as the roadbed split in two, fracturing in half a dozen other places. The bridge was a complete wreck resting on the protection pier. Meyers's timeliness and resourcefulness avoided creating a major blockade to river navigation.

With plans already in place to begin a new bridge, removal of the old bridge's remains began on July 1, 1908. A temporary bridge was not provided, closing this crossing to wagon and foot traffic until early 1910.

The third Erie Street Bridge was the last first-generation Chicago-type bascule. The design was a significant departure from the previous nine, using a strut operating mechanism instead of

Looking south at the third Erie Street Bridge in 1969. Courtesy of CDOT.

## Third Erie Street Bridge

| Opened | Bridge type | Designed by | Constructed by | Cost | Status |
|--------|-------------|-------------|----------------|------|--------|
| Feb. 1910 | Chicago type, double-leaf bascule, steel, electric powered | Division of Bridges and Viaducts, City of Chicago | King Bridge Co. (superstructure), FitzSimons & Connell Co. (substructure) | $216,000 | Removal began on Nov. 8, 1971 |

the high, curved rack-and-pinion design of earlier bridges. This system allowed for a lower, less visually obstructive superstructure and required only two trusses instead of three to support the roadway.

In 1928 the bridge's Hi-Lo laminated floor, consisting of alternating two-by-six-inch and two-by-eight-inch beams laid on end and covered with asphalt, was replaced. The new floor used six-by-twelve-inch fir subplanks, overlaid by one-by-six-inch tongue-and-grooved oak, treated with zinc chloride, and topped with two-by-eight-inch asphalt planks laid at a forty-five-degree angle. This new flooring had been first tested on a portion of the Kinzie Street Bridge, with good results. Alundum was then applied as an abrasive at a quarter-inch per pound per square foot to create a nonskid wear surface. The new lighter deck allowed for the removal of thirty tons of counterweights from the bridge.

The broken iron castings at the floor breaks were replaced with cast steel, and the stringers down the center of the roadway were replaced with steel I-beams. In 1937 the floor breaks were again replaced, new subplanks and asphalt planking were installed, and the east bridge approach received a waterproofed, reinforced concrete roadway.

Once an important connection between the North and West sides, the introduction of the Ohio Street Bridge and feeder ramp one block south in 1961 reduced traffic enough at Erie Street as to make it all but superfluous. By November 1971, Erie Street's sixty-one-year-old structure was in need of "complete rehabilitation" and was instead removed.[2] The razing of this bridge explains Erie Street's abrupt end at the east bank of the river; a new city park, which opened in 2008, however, softens the effect on the landscape.

# CHICAGO AVENUE BRIDGES

LOCATION: 640 West, 800 North; Chicago Avenue runs east and west, crossing the North Branch two miles from the river mouth at Lake Michigan.

The fourth Chicago Avenue Bridge is a good example of a second-generation Chicago-type bascule bridge influenced by the City Beautiful movement. This bridge is a single-deck, double-leaf bascule design that utilizes the internal rack and pinion patented by engineer Alexander von Babo in 1911. As explained in the patent, placing the rack and pinion within the truss steelwork, combined with a new steel girder arrangement that supported the bridge trunnions, allowed for less expensive counterweight materials, permitted better placement of the operating machinery, and left room for a gear train next to the moveable truss. Built within the same period, the Chicago Avenue, Washington Street, 92nd Avenue, and Grand Avenue bridges all shared the same gearing, rack dimensions, pitch radius, motor, and drivetrain arrangements. Duplicate gearing and operating systems were incorporated as part of the standard design to allow for continued operation in the event that one unit failed. Although the bridge spans and leaf weights differed, each of these bridges was driven by a direct current motor of 40 to 50 horsepower.

The most apparent variation from similar bridges of the second-generation Chicago type is the oval-shaped bridge houses. Chicago Avenue Bridge was the first in which the masonry and stonework of the approach and foundations integrated the bridge house into the overall structure. Previously, the bridge houses, usually made of wood, were supported by steelwork that extended off the side of the bridge approach. Unlike most of the other bridges built during this period, architect Edward Bennett did not formally review the plans for this bridge; yet, clearly, the general influences demanding better bridge architecture are evident.

The roadway of the Chicago Avenue Bridge, designed to support mostly horse-drawn wagons and light automobile traffic, was paved with creosote wood blocks. Twenty years later, this bridge was redecked. The City furnished the materials, and work was done under contract by the Great Lakes Dredge & Dock Company. New streetcar stringers and supports were added along with a timber deck of six-by-twelve-inch subplanks to support solid three-by-six-inch Douglas-fir planking, covered by a wearing surface of one-and-a-half-inch asphalt planks. The

| Current (fourth) Chicago Avenue Bridge | | | | | |
|---|---|---|---|---|---|
| Opened | Bridge type | Designed by | Constructed by | Cost | Status |
| Oct. 15, 1914 | Chicago type, double-leaf bascule, steel, electric powered | Division of Bridges and Viaducts, City of Chicago | Ketler-Elliott Co. (superstructure), Byrne Bros. Dredging & Excavation Co. (substructure) | $294,827 | Currently in use |

Looking north at the current Chicago Avenue Bridge in 2007 with the former Montgomery Ward building in the background. © Laura Banick

## First Chicago Avenue Bridge

| Opened | Bridge type | Designed by | Constructed by | Cost | Status |
|--------|-------------|-------------|----------------|------|--------|
| 1849 | Wood bent with a twenty-foot draw | Unknown | Unknown | Unknown | Removed in 1867 |

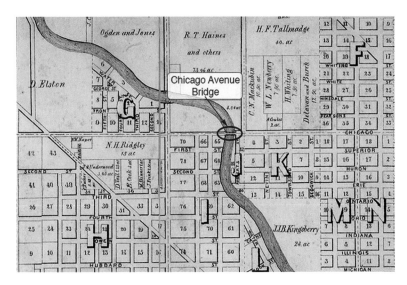

Map showing the first Chicago Avenue Bridge in 1849.

a fixed span, and the leaves are locked together by I-beams at the center of the bridge.

The exact origin of the first Chicago Avenue Bridge is unknown; however, according to the Rees and Rucker "Map of Chicago and Vicinity" of 1849, it was likely built after the flood of 1849. This bridge probably replaced two older bridges nearby at Union and Pearson streets.

A petition to the Common Council in July 1850 stated that the bridge's existing twenty-foot draw was insufficient for "vessels of ordinary size" to pass through and requested that an additional twenty feet be constructed and incorporated into the bridge, but the petition was tabled.[1] In March George W. Rodgers was appointed as bridge tender in charge of both the Kinzie Street and the Chicago Avenue bridges.

The bridge was threatened on February 8, 1857, by an early thaw and rains. A wall of ice and debris formed against the bridge, and the water level rose four feet, flooding the streets and low ground for several blocks. The immense pressure was

new deck, completed in 1932, required the addition of fifty tons of counterweights to balance the bridge. Additional redecking and major repairs to the bridge occurred in 1968 and 1969, and again in 1992. Like its neighbors, the structure now operates like

## Second Chicago Avenue Bridge

| Opened | Bridge type | Designed by | Constructed by | Cost | Status |
|--------|-------------|-------------|----------------|------|--------|
| 1867 | Swing, wood and iron, hand operated | Fox & Howard | Fox & Howard | $26,700 | Destroyed by fire on Oct. 8–10, 1871 |

## Third Chicago Avenue Bridge

| Opened | Bridge type | Designed by | Constructed by | Cost | Status |
|---|---|---|---|---|---|
| 1872 | Swing, wood and iron, hand operated | Fox & Howard | Fox & Howard | $20,850 | Replaced by a temporary pontoon bridge in 1911 |

finally released when the bridge could finally be thrown open later that afternoon and the ice allowed to move downstream. Ten years later, the bridge was torn down and replaced.

The second Chicago Avenue Bridge, built in 1867, was a modern swing bridge. It was 175 feet long and constructed of wood braces and iron chords utilizing the Howe-truss and turntable improvements patented by J. K. Thompson. It was destroyed four years later in the Great Fire of 1871.

By 1872 a new swing bridge was in operation at Chicago Avenue and was of the same length and construction as the previous bridge except for a new stone center pier. In 1903 the western approach of this bridge was completely rebuilt, as the old timber and pilings supporting it had decayed; the bridge seats were also rebuilt, and several web members in the trusses were replaced. In 1908 the turntable track was leveled; the entire bridge deck, subplanking, and deck of the approaches were replaced; the bottom chords were packed and reinforced; and the lateral braces were replaced. On January 28, 1911, the secretary of war ordered removal of this bridge and replacement with a vertical-lift or bascule bridge by early 1914. Removal of the Metropolitan West Side Elevated Train Bridge and the Jackson Boulevard Bridge was also cited in this order aimed at establishing "a broad and unobstructed navigable channel in the Chicago river and its branches."[2]

A temporary pontoon bridge was built at a cost of $28,512.80 by the National Contracting Company to carry pedestrians and opened on January 21, 1912. The FitzSimons and Connell Company removed the old bridge, center pier, and protection pier between February and March 1912 for $800.

# PEARSON AND UNION STREET BRIDGES

(No Current Bridges)

LOCATION: 640 West, 820 North; Pearson Street runs east and west, but no longer reaches the river and dead-ends just west of State Street. Union Street, running north and south, now runs only two short blocks near the North Branch between Hubbard and Ohio streets. The two streets both crossed either end of a sharp bend in the North Branch.

Two fixed, most likely wood bent bridges crossed the North Branch at Union and Pearson streets near Huntoon's Sawmill (noted on the map). It is uncertain when these two bridges were built or removed, and the only documentation of their ever having existed is shown in the excerpt from a Canal Commission map from 1836. The sawmill was built by Mark Nobel in 1832; it burned in 1834. Gurdon S. Hubbard became the new owner, and the mill was refitted and leased by Captain Bensley Huntoon, with whom it became associated. The sawmill was closed in 1840, but would next be owned by the Crosby family and expanded into a distillery in the 1850s. The two bridges at Union and Pearson would have served the mill and were likely constructed and maintained by the proprietors. Such wood

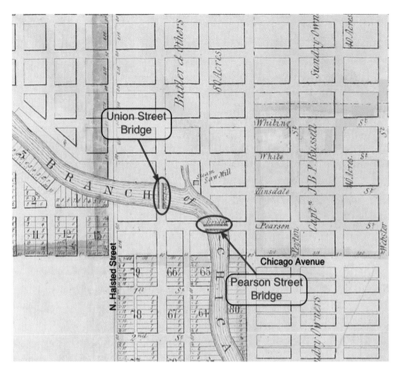

Map showing Union Street Bridge, Pearson Street Bridge, and Hooton's Sawmill in 1836.

bridges would typically have needed refurbishing, if not replacing, every four to five years and certainly would have been removed after the improvements made to the North Branch for ships and the construction of the first Chicago Avenue Bridge in 1849.

## Pearson & Union Street Bridges

| Opened | Bridge type | Designed by | Constructed by | Cost | Status |
|---|---|---|---|---|---|
| Pre-1836 | Fixed, wood bent | Unknown | Unknown | Unknown | Removed in the 1840s |

# NORTH HALSTED STREET BRIDGES (RIVER)

LOCATION: 600 West, 1047 North; Halsted Street runs north and south, crossing the North Branch 2.6 miles from the river mouth at Lake Michigan.

The fourth and current North Halsted Street Bridge is a modern double-leaf Chicago-type bascule bridge. Built in 1955, this deck-truss bridge has a pleasing profile, with a roadbed supported entirely from below by arched steel trusswork. It was the first bascule bridge in Chicago built with a single bridge house and designed for one-person operation. It measures 184½ feet between the trunnions, and its 46-foot-wide open steel-grid roadway, which was refurbished in 2012, carries four lanes of traffic. The structure now acts as a fixed bridge, as huge clamped cables, with multiple wraps, lock the largest operating gear permanently in place on both sets of operating machinery, preventing its operation as a drawbridge.

The first bridge at North Halsted Street was financed through twenty-five hundred dollars in subscriptions from local property owners and forty-five hundred dollars from the City of Chicago. This 140'-long × 20'-wide all-wood swing bridge was constructed with a center pier supported by timber pilings and approaches at each riverbank.

Just east of this North Halsted Avenue Bridge, a key site in the city's shipbuilding industry was established on what would become part of Goose Island. George Wicks started a shipyard in 1854 by constructing the city's first dry dock on the east bank of the river below the North Halsted Avenue Bridge. Before it was finished, he sold his interest to Doolittle & Miller, who completed the 275'-long × 8'-deep dry dock in 1855. The firm Doolittle & Miller had already built facilities on the river in 1848–49 to support the construction of three- to four-hundred-ton ships on the South Branch near Van Buren Street.

With the completion of the North Branch Canal in the 1850s, the site was situated at the southern tip of Goose Island. In 1871–72 a second dry dock was added that admitted vessels up to 310 feet in length. The firm now had the largest dry docks on Lake Michigan; a third dry dock was added in 1885. By then named Miller Brothers & Company, the firm would build steamships, tugs, canal boats, and lake schooners to become the largest and most important shipbuilder in Chicago up through the early twentieth century. In 1900 the firm was sold to the Ship Owner's Dry Dock Company and in 1907 sold again to the

| Current (fourth) North Halsted Street Bridge (River) | | | | | |
|---|---|---|---|---|---|
| Opened | Bridge type | Designed by | Constructed by | Cost | Status |
| Oct. 5, 1955 | Chicago type, double-leaf bascule, steel, electric powered | Division of Bridges and Viaducts, City of Chicago (substructure) | Overland Construction Co. (superstructure), Kenny Construction Co. | $3.3 million | Currently in use |

Looking north at the current North Halsted Street Bridge over the river in 2007. © Laura Banick

Chicago Ship Building Company, which was then acquired by the Chicago subsidiary of the American Ship Building Company. The shipyard produced vessels that served in the Civil War and World War I. In 1916 the Goose Island facility was closed, and the three dry docks were filled in during construction of the Ogden Avenue Bridge and overpass in the late 1920s.

In 1877 the Department of Public Works reported that the "North Halsted and North Avenue bridges cannot be depended on with any degree of certainty for any length of time after the opening of navigation, as the danger is not so much while in position for travel, as it is of breaking down while being swung."[1] Later that year the North Halsted Street Bridge was removed.

On June 22, 1877, a contract for a second North Halstead Bridge was extended to William B. Howard to erect a combination wood and iron drawbridge over the North Branch at Halsted Street, which was completed by September 1, 1877. This second swing bridge at North Halsted Street was of more substantial construction than its predecessor, although both were hand operated.

By 1893 the Department of Public Works reported that this wood-truss bridge had been rebuilt and patched to such an

## First North Halsted Street Bridge (River)

| Opened | Bridge type | Designed by | Constructed by | Cost | Status |
|---|---|---|---|---|---|
| 1866 | Swing, wood, hand operated | Fox & Howard | Fox & Howard | $7,000 | Removed in 1877 |

extent that the only original elements were the iron chords, and those were much deteriorated from the sulfuric acid as a result of the coal and tugboat fumes. After serving for nineteen years, the second North Halsted Street Bridge was removed in 1896.

For the third North Halstead Street Bridge, the City initially favored a swing bridge to replace the old wood and iron bridge; marine interests got wind of the City's plans, however, and insisted that a bascule or vertical-lift bridge be built. North Halsted Street received a Scherzer rolling-lift bascule as its third bridge, only the third such bridge built in the city. The first two Scherzer bridges offered an eighty-five-foot river channel, while this larger Scherzer rolling lift provided a one-hundred-foot waterway for navigation. The double-leaf bridge had a span of thirty-four feet between the two supporting trusses that carried two sets of trolley tracks down its center and wagon teams on either side. Cantilevered over the river were six-foot-wide sidewalks on each side. In 1910 the bridge was redecked following a widening of the streetcar

rails to accommodate the new "pay as you enter" streetcars.[2] It was also repaved with new oak planking in 1912. Eleven years later, extensive structural, floor-system, soleplate, track-plate, and operating-machinery repairs were also completed.

On September 11, 1913, the bridge was the scene of the disappearance of salesman Rocco Nigro. The automobile in which he was being chauffeured, a Locomobile Tonneau, was heading southbound toward the bridge "at a high rate of speed" just as it was opening.[3] The driver, a man named Bakes, neither heard nor saw the bells or signal as he drove around a wagon stopped at the approach. With no warning gates on the bridge, Bakes realized the danger only after the car dropped about a foot and a half onto the rising bridge deck. As the car raced up the incline, Bakes began to furiously work the brakes, bringing the car to a sudden stop at the brink with the front wheels hanging over the edge. When Bakes turned to look in the backseat, Nigro was gone.

## Second North Halsted Street Bridge (River)

| Opened | Bridge type | Designed by | Constructed by | Cost | Status |
|---|---|---|---|---|---|
| Sept. 1, 1877 | Swing, wood and iron, hand operated | William B. Howard | W. B. Howard (superstructure), Chicago Dredge & Dock Co. (substructure) | $14,680 | Removed in 1896 |

## Third North Halsted Street Bridge (River)

| Opened | Bridge type | Designed by | Constructed by | Cost | Status |
|---|---|---|---|---|---|
| Jan. 19, 1897 | Scherzer rolling lift, double-leaf bascule, steel, electric powered | Scherzer Rolling Lift Bridge Co. | King Bridge Co. (superstructure), Wilson & Jackson (substructure) | $114,000 | Removed on March 1, 1955 |

Looking north at the open third North Halsted Street Bridge in 1908.

A 1909 Locomobile 40 Baby Tonneau automobile.

**NO ONE HEARD ANY SPLASH**

*Chicago Tribune* cartoon of the incident in 1913.

Bakes explained to the police, "I don't know what happened to Nigro, or where he went. I didn't see him fall into the river and I don't see how he could have. I was sitting in the front and he was sitting in the rear. I was talking to him just as we went onto the bridge. When we came to a stop up there in the air I looked around and saw his hat on the seat—that's all. I don't know if he was drowned. He may have run away, fearing arrest."[4]

In 1952 the City began construction of a fourth North Halsted Street Bridge. It was designed in such a way that the new structure could be built around the old bridge and kept in service for as long as possible. In March 1955, this old bridge was removed. The new bridge was lowered and reopened to street traffic in December 1955.

# NORTH HALSTED STREET BRIDGES (CANAL)

LOCATION: 800 West, 847 North; Halsted Street runs north and south, crossing the canal on the North Branch 2.8 miles from the river mouth at Lake Michigan.

This fixed-arch suspension bridge was constructed after the removal of a 102-year-old Chicago-type bascule bridge in 2011. The new four-lane bridge is a significant improvement over the old two-lane bridge it replaced. Designed by Muller+Muller Limited of Chicago, the bridge also features marked bike lanes that connect with the existing street bicycle lanes to the north and south of the bridge.

The first North Halsted Avenue Bridge was built on swampy, undeveloped land first granted to the Illinois & Michigan Canal Commission by the federal government in the 1830s. Surveyed in 1836, it was sold to raise funds for the digging of the I&M Canal. The addition of the North Branch Canal to make the land more attractive to the city's growing industry created the need for a bridge there. The sand and clay collected from digging the canal were used to make bricks, and the remaining tailings were used to raise the grade of the surrounding area.

A major street for northbound and southbound traffic, Halsted Street was reconnected with the first North Halsted Avenue Bridge in 1874. It was only the second bridge to cross the North Branch Canal. This swing bridge was planned by the Chicago Department of Public Works in the early 1870s. City engineers faced several challenges in designing the bridge. A major street, Halsted conformed to a cardinal-point grid, and the street plan required maintaining this alignment. The northwest-to-southeast course of the canal meant the bridge would cross the waterway on a diagonal. The shortest span and most economical solution would have been a fixed bridge with a skewed-truss structure matching the banks of the canal; however, a skewed-truss design was not readily adaptable to a swing bridge, and the waterway required a moveable bridge. The solution came in the form of the 226-foot North Halsted Street Bridge. This iron swing span rotated on a center pier constructed on the east bank of the canal, and the entire eastern half of the bridge extended entirely over land.

This bridge received a new brick bridge house in 1875. Around the same time, a double gear was attached to both of the pinions to facilitate operating the bridge in stormy weather. This upgrade made it possible for two bridge tenders to control

| Current (third) North Halsted Street Bridge (Canal) | | | | | |
| --- | --- | --- | --- | --- | --- |
| Opened | Bridge type | Designed by | Constructed by | Cost | Status |
| 2012 | Fixed, arch suspension | Division of Bridges and Viaducts, City of Chicago, and architects Muller+Muller, Ltd. | Walsh Construction Co. | $13,097,568 | Curently in use |

Looking south at the current North Halsted Street Bridge over the North Branch Canal in 2012. © Patrick McBriarty

Looking south at the first North Halsted Street bridge over the North Branch Canal in 1906. Courtesy of the Chicago History Museum.

## First North Halsted Street Bridge (Canal)

| Opened | Bridge type | Designed by | Constructed by | Cost | Status |
|---|---|---|---|---|---|
| 1874 | Swing, iron, hand operated | Fox & Howard | Fox & Howard | $29,945 | Removed in Feb. 1907 |

the bridge even on particularly windy days. In 1887 the bridge was raised so the timber planking supporting the turntable could be replaced, beneath which seven chords of stone and gravel were added to strengthen the center pier. A year later, both of the approaches were replanked; the bridge stringers, sidewalks, and railings were replaced; and street-railway tracks were installed.

By 1905 the Bridge Department reported "considerable difficulties" keeping this bridge in service throughout the year, as the ironwork was "very badly corroded and almost constant patching was required" to keep the structure safe.[1] On June 27,

1906, the poor condition of this bridge necessitated prohibition of team and streetcar traffic until February 21, 1907, when foot traffic was also halted and the bridge's removal commenced.

This second North Halstead Street Bridge was a first-generation Chicago-type bascule completed in 1908. The 206'-long × 60'-wide bridge carried both streetcar and vehicle traffic until the streetcar tracks were removed in the 1950s with the City's shift to buses.

In 1911 the roadway was paved with sectional wood pavement placed outside the streetcar tracks. The approach pavement

## Second North Halsted Street Bridge (Canal)

| Opened | Bridge type | Designed by | Constructed by | Cost | Status |
|---|---|---|---|---|---|
| Nov. 4, 1908 | Chicago type, double-leaf bascule, steel, electric powered | Division of Bridges and Viaducts, City of Chicago | J. E. Roemheld (superstructure), FitzSimons & Connell Co. (substructure) | $247,983 | Removed in Dec. 2010 |

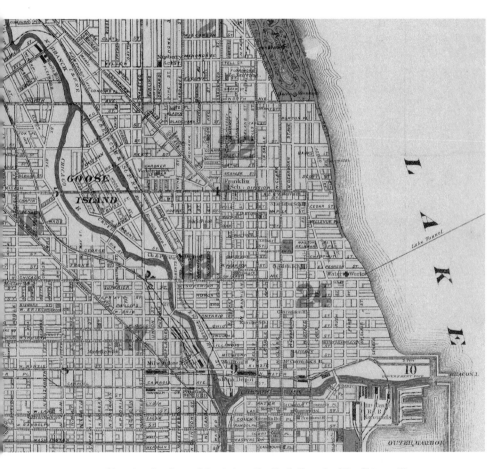

Map showing Goose Island along the North Branch of the Chicago River, which was created in 1857 with the addition of the North Branch Canal.

was patched, electric gong and stop signals were installed, the center locks were repaired, and the bridge houses were painted. In 1916 the original North Halsted Avenue bridge houses, at the northwest and southeast corners, were completely rebuilt. The bridge roadway was redecked in 1918. In 1931 the bridge received a new floor, and several trusses were reinforced. New sidewalks and counterweight boxes were installed, requiring that an additional 130 tons be added to the counterweights. The bridge was painted and received new center locks and electric roadway gates with interlocks. In 1955 this bridge was resurfaced with an open steel-grid decking.

In the mid-1990s, the bridge was converted from a moveable span to a fixed span. The bridge houses, made obsolete by the conversion, were removed. In 2009 the bridge received a new open-grate bridge deck; in 2010 the bridge was closed to traffic and then removed for replacement by a new fixed bridge.

Looking south at the second North Halsted Street Bridge over the North Branch Canal in 2007. © Laura Banick

# OGDEN AVENUE BRIDGES (RIVER AND CANAL)

(No Current Bridges)

LOCATION: 950 West, 965 North (river), and 900 West, 1136 North (canal). There are no Ogden Avenue bridges left today. This southwest-northeast boulevard first crossed the North Branch of the river and the North Branch Canal 2.9 miles from the river mouth.

HISTORICAL HIGHLIGHT: Originally called the Southwest Plank Road, this street was renamed Ogden Avenue in 1878 after the death of William B. Ogden, former mayor and longtime booster of the City of Chicago.

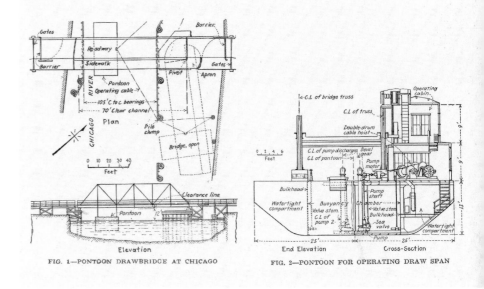

FIG. 1—PONTOON DRAWBRIDGE AT CHICAGO        FIG. 2—PONTOON FOR OPERATING DRAW SPAN

Design drawings of the first Ogden Avenue pontoon swing bridge over the river in 1908.

The extension of Ogden Avenue east across the river to Lincoln Park was first seriously proposed in the *Plan of Chicago* of 1909, though the idea initially arose during several aldermanic races in the 1880s. In 1922 the City began the project, announcing it would tear down buildings to create a 108-foot-wide highway running from Union Park (at Ogden Avenue and Lake Street) to Lincoln Park. In 1924 a temporary pontoon swing bridge was installed across the river at Ogden Avenue. Plans for two bascule bridges had been completed and approved. The plan was to replace this temporary bridge once funds were made available.

Reusing a steel thru-truss, the 105-foot draw of the bridge was connected to the shore by timber trestle approaches. At one end, a corner was supported by a ball and socket, while the other corner rested on a roller carriage and curved track. The opposite, free, end of the draw rested on a skid timber between the bridge and a bent-pier approach. Beneath this free end was a steel pontoon 50' × 22' × 13' deep. The pontoon held the bridge operating machinery and was rigged with pumps and sea valves.

| First Ogden Avenue Bridge | | | | | |
| --- | --- | --- | --- | --- | --- |
| Opened | Bridge type | Designed by | Constructed by | Cost | Status |
| Nov. 4, 1908 | Pontoon swing, steel, electric powered | Division of Bridges and Viaducts, City of Chicago | City day labor | $200,000 | Removed in 1933 |

While in the closed position, the pontoon was partially flooded to hold it in position; in order to open it, water was pumped out, increasing its buoyancy and raising the bridge off the approach and thereby freeing it to rotate. A 50-horsepower hoist motor turned double drums controlling two ¾" steel cables. One cable was anchored to a piling cluster above and the other to pilings below the bridge, so that, as one cable slackened, the other was pulled by the drum to open the bridge. The motor was reversed to close the bridge, and the partially flooded pontoon sank into place before restoring street traffic.

In 1931 Chicago's Board of Local Improvements removed this temporary pontoon bridge, and the Division of Bridges and Viaducts stored the dismantled bridge at the municipal plant. This made way for the construction of the Ogden Avenue Overpass.

These two bascule overpass bridges opened in 1932, crossing the river and the North Branch Canal to extend Ogden Avenue east to Lincoln Park. The 3,400-foot overpass was part of a fourteen-year, twenty-million-dollar Ogden Avenue widening and improvement project. The overpass brought Illinois State Route 34 over the river to the intersection of North Avenue and Larrabee Street. The Board of Local Improvements oversaw the entire project. The designer of these bridges is unknown, and there is no mention of the work being done by the City Bridge Department throughout the 1920s or early 1930s.

In hindsight, poor planning and inadequate links to major traffic arteries like Lake Shore Drive or the Eisenhower Expressway meant the expected traffic on the overpass never materialized. Lack of planning, use of street frontage, and poor commercial development also contributed to the project's failure. In addition, the Interstate Highway System constructed during the 1960s never tied into Ogden Avenue.

In 1960 these bridges were redecked with open steel grating. In 1993 both bridges were closed to traffic; the Paschen Construction Company removed the entire overpass structure, including the drawbridges, in 1994. Illinois State Route 34 was truncated at Roosevelt Road, and now only one block of Ogden Avenue remains east of the river. The abandoned land was opened to public parks and real-estate development.

| Second Ogden Avenue Bridge (River) and first Ogden Avenue Bridge (Canal) | | | | | |
| --- | --- | --- | --- | --- | --- |
| Opened | Bridge type | Designed by | Constructed by | Cost | Status |
| Dec. 9, 1932 | Chicago type, double-leaf bascule, steel, electric powered | Unknown | Great Lakes Dredge & Dock Co. | $4.6 million | Closed in 1993 and removed in 1994 |

Aerial view looking northeast during construction of the Ogden Avenue Overpass with raised bascule bridges over the river and the canal. Courtesy of CDOT.

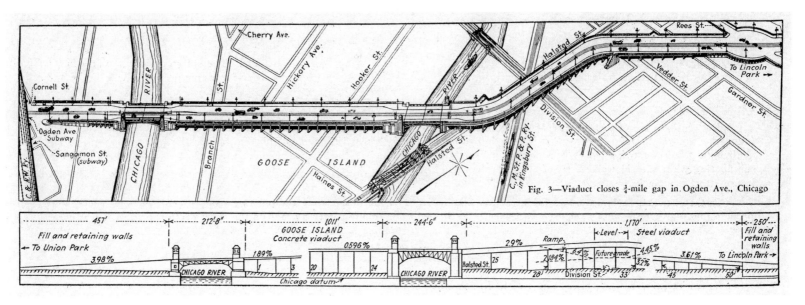

Drawing of aerial and side view north of the new Ogden Overpass from 1931. Courtesy of CDOT.

# WEST DIVISION STREET BRIDGES (RIVER)

LOCATION: 1129 West, 1200 North; Division Street runs east and west, connecting with the west side of Goose Island and crossing the river over the North Branch three miles from the river mouth at Lake Michigan.

The current West Division Street Bridge was opened in 1904 on the heels of the new East Division Street Bridge being built over the North Branch Canal. These two bridges connected Goose Island with the West and North sides. The West Division Street Bridge was the fourth Chicago-type bascule ever built. A first-generation double-leaf trunnion bascule, it used seven hundred tons of structural steel and five hundred tons of counterweight material. The length between the trunnions was slightly less than 173 feet, and the three-truss structure supported two 18-foot-wide roadways and two 8-foot-wide sidewalks cantilevered off each side of the bridge. The same year it was built, new larger brakes were added to the bridge, as the originals were found to be inadequate. Pit pumps were installed that reused electric motors from Canal Street, and substantial rewiring and alterations were done to the switchboards. The bridge houses were also painted.

This century-old bridge has undergone a significant number of repairs. In 1907 the roadway was paved between the outer streetcar rails and the wheel guards, the ironwork was painted, and a new and improved danger-signal system was installed. In the 1930s, the bridge houses were rebuilt, and the bridge received new center locks, machinery renewal, and deck replacement. In 1942 the bridge was again redecked and received a number of structural steel repairs. The rear end of each bridge leaf received seven feet of concrete slabs, allowing the removal of approximately twelve tons from the counterweight of each leaf. In the mid-1950s, the plank roadway on the end of the moveable leaves was replaced with an open steel-grid deck. In the late 1960s, the original operator houses were replaced with modernist structures sporting flat roofs, overhanging eaves, and vertical wood siding.

The Department of Public Works considered replacing the bridge in the 1980s with a new double-leaf bascule; though plans were prepared, construction was never funded. Instead, the West Division Street Bridge was almost completely rebuilt with new structural steel in 1993. The span's general appearance was preserved, although the new trusses differ slightly from the original pattern, and the steelwork was bolted together rather than riveted. This project also converted the bridge to a fixed

| **Current (second) Division Street Bridge (River)** | | | | | |
|---|---|---|---|---|---|
| Opened | Bridge type | Designed by | Constructed by | Cost | Status |
| June 4, 1904 | Chicago type, double-leaf bascule, steel, electric powered | Division of Bridges and Viaducts, City of Chicago | J. E. Roemheld (superstructure), FitzSimons & Connell Co. (substructure) | $256,315 | Currently in use |

Current West Division Street Bridge over the river, looking north in 2007. © Laura Banick

The first West Division Street Bridge in 1902, looking north.

span; the bridge houses were cleared of all control equipment and the upper buffers removed, while the trunnions, racks, and drivetrains remained. The lower buffers were locked in place, and the movable leaves were bolted together.

The first West Division Street Bridge was built after the Civil War, as Chicago continued to grow as a major commercial and industrial center, and the North Branch Corridor became a strong

manufacturing and industrial area. The development began soon after the canal was completed in the late 1860s. Construction of this first West Division Street swing bridge and its eastern counterpart over the canal encouraged further development of Goose Island. The West Division Street Bridge was hand operated and was referred to as a combination swing bridge, as the 180'-long × 29'-wide Howe-truss bridge was made of

## First Division Street Bridge (River)

| Opened | Bridge type | Designed by | Constructed by | Cost | Status |
|---|---|---|---|---|---|
| 1869 | Swing, wood and iron, hand operated | Fox & Howard | Fox & Howard | $15,795 | Removed on May 1, 1902, and reused at Blackhawk Street |

wood and iron. Iron compression rods ran vertically and bolted top and bottom to the wood upper and lower truss beams, also known as chords. Crossing wood members created the truss structure, and overhead crossbeams connected the two trusses over the center of the bridge. A small bridge house was built off one side at the midpoint of the bridge for the storage of tools and protection of the bridge tenders during inclement weather.

In 1870 a Board of Public Works traffic survey conducted during an average thirteen-hour day recorded 900 pedestrians and 130 vehicles crossing the Division Street Bridge, making it one of the least traveled of the twenty-seven City bridges. In 1895 a new floor system was installed to support electric-streetcar use. An 1898 survey by the City Bridge Department reported that the bridge's wood supporting members were rotting away, the ironwork was badly rusted, and the center pier was shaky and rotten. The bridge was removed four years later once financing was made available, and the bridge superstructure was reused at Blackhawk Street.

# EAST DIVISION STREET BRIDGES (CANAL)

LOCATION: 840 West, 1200 North; Division Street runs east and west, crossing the North Branch Canal to connect to the eastern side of Goose Island with the North Side 2.6 miles from the river mouth.

A first-generation Chicago-type bascule, the second Division Street Bridge is the same general design as the 95th Street Bridge that was built in 1903 and removed in 1958. The East Division Street Bridge was the second Chicago-type bascule ever built. It measures 146 feet between the trunnions and 60 feet wide and opened to traffic on February 1, 1903. The overhead cross-bracing of this bridge is adorned with a series of Ys, long a representative symbol of Chicago relating to the Y-shaped river that divides the city. The bridge opened with little or no fanfare, as the West Division Street Bridge had been removed six months earlier. Local citizens expressed frustration at the City's poor coordination both in replacing these bridges and in extending the closure of the only east-west street running across Goose Island. The street remained severed until June 1904, when the West Division Street Bridge opened.

The designs for this bridge and the 95th Street Bridge most closely resembled the initial designs by City engineers for the Chicago-type bridge. Both bridges were put out to bid within two weeks of each other on May 15 and June 1, 1900. The firm of Roemheld and Gallery was the low bidder in each case and constructed both bridges based on the municipal designs. Separately, three independent consulting engineers reviewed the City designs and recommended several changes, which were later incorporated into the Courtland Street Bridge.

At Division and 95th Streets, the counterweight pit was split in-line with the center truss to create two side-by-side pits. The abutment was also somewhat higher than the river pier. At Courtland Street and subsequent first-generation Chicago-type bridges, the abutments were lowered to match the height of the river pier. Also, a single large counterweight pit was used that allowed the ends of the trusses to be tied together for greater rigidity. It is believed that the Division and 95th Street bridges did incorporate one recommendation by adding steel reinforcement to strengthen the abutments and river piers. Unexpected site problems and changes caused both bridges to run about 50 percent over bid, which prompted a public outcry. Officials investigated both projects for fraud but were unable to find any clear wrongdoing.

The Division Street Bridge received its first major rehabilitation in the early 1930s, when the steel superstructure was reinforced and the counterweights, trunnion trusses, and bridge

| Current (second) Division Street Bridge (Canal) | | | | | |
| --- | --- | --- | --- | --- | --- |
| Opened | Bridge type | Designed by | Constructed by | Cost | Status |
| Feb. 1, 1903 | Chicago type, double-leaf bascule, steel, electric powered | Division of Bridges and Viaducts, City of Chicago | Roemheld & Gallery | $194,150 | Currently in use |

Looking north at the current Division Street Bridge over the canal in 2007. © Laura Banick

deck were rebuilt. In 1942 this bridge was redecked, received a number of structural steel repairs, and, like the West Division Street Bridge, had concrete added to the rear decks to allow for the removal of approximately twelve tons of counterweights from each leaf. A similar overhaul was conducted in 1969, and extensive rehabilitation of the superstructure was again completed in the early 1980s. The East Division Street Bridge was converted to a fixed span in the early 1990s. The Division Street Bridges are two of the four remaining first-generation Chicago-type bascule bridges.

In 1870 this East Division Street Bridge was the first crossing built over the North Branch Canal. It was a 176'-long × 29'-wide hand-operated wood and iron bridge. The firm of Fox & Howard, the contractors for the excavation of the canal, also built the

## First Division Street Bridge (Canal)

| Opened | Bridge type | Designed by | Constructed by | Cost | Status |
|--------|-------------|-------------|----------------|------|--------|
| 1870 | Swing, wood and iron, hand operated | Fox & Howard | Fox & Howard | $17,700 | Removed in Aug. 1900 |

bridge. It was Harry Fox, in partnership with John P. Chapin, who dredged much of the forks of the Chicago River and first suggested using the fill to raise the grade of the streets in the 1850s. Then, from 1860 to 1875, he partnered with William B. Howard, a practical bridge builder, to build twenty-six swing bridges and one fixed bridge over the Chicago River. Their firm would also significantly dredge the waterways and build nearly fifteen miles of docking along the river. The firm was awarded far-ranging projects throughout the South and Midwest and employed as many as six to eight hundred people. Harry Fox would work with the railroads and later became a partner in the FitzSimons and Connell Company, which specialized in dredge, dock, and bridge construction projects. In 1895 this bridge received a new floor to support its use by electric streetcars.

On a Thursday evening, March 9, 1899, fifty passengers "packed like sardines" into a streetcar were nearly dumped into the river at this bridge.[1] The trolley driver, John Mullens, did not notice that the bridge was open and drove the heavy trolley car toward the opening until the very last moment. Despite his at-tempts to brake, the front wheels dropped off the tracks down onto the bridge abutment two feet below. The sudden stop shattered the glass of the windows and threw several passengers to the floor of the car. Mullens screamed for the passengers to jump for their lives and leaped into the dirty river. A wild scramble ensued in which four passengers were injured, though not seriously. Mullens, who was pulled from the river by the bridge tender, declared between gasps that he was not responsible for the accident; witnesses, however, countered that the streetcar was running much too fast. In anticipation of such a situation, the trolley wires were designed so that within one hundred feet of the river, the power would shut off whenever the bridge was turned; however, the excessive speed of the streetcar had nearly overcome this safety measure.

The bridge was removed a year later, in 1900. While a new bascule bridge was being constructed, the streetcars were re-routed, and a temporary pontoon bridge carried pedestrian traffic at this location.

# BLACKHAWK STREET BRIDGE

(No Current Bridge)

LOCATION: 1000 West, 1500 North; Blackhawk Street runs east and west, and though there is no bridge at this location today, it did have a bridge over the North Branch Canal three miles upstream of the river mouth between 1902 and 1910.

On May 15, 1902, Blackhawk Street received a temporary bridge providing access to Goose Island while a new bridge was being built at West Division Street. The Blackhawk Street Bridge used the old superstructure of the West Division Street Bridge. It was repaired, strengthened, and placed on a new center pier. The FitzSimons and Connell Company built the substructure, while City day laborers repaired the sidewalks, fences, roadway, and chord covering; overhauled the turntable spider rods, drum, rack, and locks; and resheeted the bridge house. In May 1902, the 180'-long × 29'-wide swing bridge was opened to traffic.

Chord covering, usually made from tin or iron, was used to preserve the wood chords of bridge trusses; otherwise, in most climates, dry rot regularly occurs in untreated wood within four to five years when exposed to the elements. The problem with dry rot is that it weakens wood beams from the inside out, especially dangerous because it is often undetectable from a surface inspection. Chord covering was one solution; however, it too was subject to deterioration or corrosion, particularly from sulfuric acid created by mixing sulfur and water. The burning of coal and resulting sulfur in smoke from city industries, tugboats, and steamships combined with moisture in the atmosphere to make this a protracted problem. Regular maintenance, inspection, and painting were critical in preserving the bridges of the day.

On June 2, 1904, the steamer *Black Rock* collided with the north end of the center-pier protection and shoved the open bridge about five feet south, breaking nearly all the piles in the center pier. Although the West Division Street Bridge opened to traffic two days later, the Blackhawk Street Bridge was removed

| | | | | | |
|---|---|---|---|---|---|
| **Blackhawk Street Bridge** | | | | | |
| Opened | Bridge type | Designed by | Constructed by | Cost | Status |
| May 1902 | Swing, wood and iron, hand operated | Fox & Howard | Fox & Howard (superstructure), FitzSimons & Connell Co. (substructure) | Unknown | Removed on Apr. 9, 1910 |

Looking north at the Blackhawk Street Bridge in 1902.

on scows and repaired, including completely rebuilding the center pier. It was put back in operation on December 1, 1904. In 1909 the old streetcar rails were removed, about half the floor beams were replaced, the bridge was replanked, several truss members were replaced, and the chord covering was replaced.

Then, in 1910, this bridge was permanently removed, having been declared an obstruction to navigation by an order from the U.S. secretary of war. The FitzSimons and Connell Company completed the removal for an estimated $2,012, relieving Blackhawk Street of its one and only bridge.

# WEED STREET BRIDGES

(No Current Bridge)

LOCATION: 1050 West, 1550 North; Weed Street runs east and west and has not had a bridge since 1914. This street was previously host to two bridges, crossing the North Branch Canal 3.2 miles from the river mouth.

HISTORICAL HIGHLIGHT: Weed Street was the site of the first of only three Harman folding-lift bridges ever built. (For more details, see the "Canal Street Bridges" chapter.)

The Weed Street Bridge was a first attempt by the City to find an alternative to the obstructive center-pier swing bridges. Contracted in 1890, this bridge was designed and patented by Captain William Harman, manager of the Chicago Towing Company and former blacksmith. Completed in June 1891, the 150-foot bridge with a 62-foot draw was considered experimental and "cheaply constructed."[1]

To the inventor's disgust, his folding-lift design became known as a *jackknife* bridge for its mode of operation. A counterweighted cam, connected to the lifting cables, rotated in a downward motion to aid in the raising of the bridge. It required significant force to operate and was perfectly balanced only when in the fully open position. Yet the design had several advantages.

Drawing of folding-lift bridge at Weed Street from *American Scientific* in 1891.

The bridge leaves formed a secure and firm roadway and when open presented an impassable barrier to traffic approaching the river from the street. This solved the common problem of unattended or spooked horse teams running into the river. In addition, wharves or piers on either side of the bridge could be used

| First Weed Street Bridge | | | | | |
|---|---|---|---|---|---|
| Opened | Bridge type | Designed by | Constructed by | Cost | Status |
| June 24, 1891 | Harman folding lift, iron and steel, hand operated | Captain William Harman | Shailer & Schniglau | $8,296 | Closed for use on Aug. 18, 1899, and removed in 1905 |

for docking without impeding operation of the bridge; because there was no center pier, the full channel of the river was open to navigation when the bridge was opened. Yet this was also true of most other bascule-bridge designs that would follow.

Inspection of the new bridge generated much interest and was attended by Mayor Washburne, various city engineers, and other dignitaries. To the dismay of the crowd, opening and closing the bridge, which was done manually, took quite a while. Harman and the contractors explained that, with a few improvements and the addition of a steam engine, the bridge could operate as quickly as any on the river. The final impression was that the concept seemed right, but the design definitely needed to be improved and refined. Harman assured folks that a greatly improved ninety- to hundred-foot span could be constructed sufficient to the needs of the Chicago River.

Ultimately, the folding-lift design was problematic because it had too many moving parts. In 1895 the cables were overhauled, and in 1897 new cables were installed. Years later it was reported that residents in the vicinity "took to the high grass" to avoid injury before it was opened, fearing it might fly apart during its mechanical spasms.[2] The first Weed Street Bridge was closed to traffic by City engineer Ericson as a safety measure in 1899, the day after the 95th Street swing bridge collapsed. The bridge remained out of service, and in 1905 it was removed.

The second Weed Street Bridge, installed in 1905, reused a temporary pontoon swing bridge that had formerly been employed at North Western Avenue. Once the approaches to the Weed Street Bridge were rebuilt, the temporary pontoon bridge was floated downriver and installed. This second Weed Street Bridge was also short-lived, however, and was declared an obstruction to navigation by the War Department in 1909. Subsequently, the City appropriated funds for its removal in 1909 and reuse at Wilson Avenue.

Although the John J. Gallery Company had a contract for its removal, on April 28, 1910, bridge engineer Thomas Pihlfeldt ordered City tug and barge workers to remove the Weed Street superstructure from its foundation. The Gallery Company completed the balance of the bridge's removal, but submitted a bill for the full contract amount of $1,348. This raised questions among City officials, and the new commissioner of public works, B. J. Mullaney, summoned Pihlfeldt to a hearing; Pihlfeldt explained that City removal of the bridge was necessary because the Gallery Company was unable to move the bridge immediately and there was a second contract to remove the bridge substructure by May 14 that would have been delayed. It was determined that the City work cost $99.60, which the John J. Gallery Company subtracted from their bill, resolving the matter. There has not been a Weed Street Bridge since.

## Second Weed Street Bridge

| Opened | Bridge type | Designed by | Constructed by | Cost | Status |
|---|---|---|---|---|---|
| 1905 | Pontoon swing, wood, hand operated | Division of Bridges and Viaducts, City of Chicago | Jackson & Corbett Co. | Unknown | Removed in Apr. 1910 |

# NORTH AVENUE BRIDGES

LOCATION: 1150 West, 1600 North; North Avenue runs east and west, crossing the North Branch 3.9 miles from the river mouth at Lake Michigan.

The current North Avenue span is the first suspension, cable-stayed bridge in Chicago. A temporary steel bridge just to the south maintained traffic during its construction. The northern half of the bridge opened to traffic in January 2008. The southern half of the bridge was not opened until the following May, so contractors could utilize it as a work platform for removal of the temporary bridge. Once completed, the entire bridge was opened to traffic, providing two lanes in each direction.

This fixed span replacing a first-generation Chicago-type bascule bridge has greatly relieved traffic congestion at this busy crossing. The benefit of fixed spans is their low cost of construction and minimal maintenance requirements when compared with a moveable bridge. The new bridge removed a serious bottleneck and now matches the existing street capacity. After dark, the lighting of the bridge, designed and created by the Chicago-based architectural lighting firm Schuler-Shook, brings it to life.

Efforts for the first North Avenue Bridge began in October 1856, as citizens along the North Branch Corridor petitioned for a bridge by pledging $620 in subscriptions for the installation of an old pontoon-float bridge from Randolph Street. The Common Council's Bridge Committee, in July 1858, responded that they were against adding another bridge and cited a tenfold increase in bridge maintenance costs between 1851 and 1857. They concluded that the City could not afford the responsibility of another river crossing unless local residents and business owners would pay the entire cost of the first bridge. Locals responded by depositing $1,000 with the City Treasury and guaranteed that they would pay all expenses for two years after completion of the bridge. In late 1858, the superintendent of public works was instructed by the Common Council to construct "a substantial float bridge across the north branch at North Avenue, the cost thereof not to exceed $3,000, nor less than $2,500."[1] The residents received their bridge, which served for seven years. In 1865 it "failed at the close of winter," and a contract for its replacement was issued on March 18, 1865.[2]

This second North Avenue Bridge was a 145'-long × 19'-wide all-wood swing bridge. It lasted eleven years before being replaced. A third North Avenue Bridge, built of wood and iron and measuring 150 feet long by 29 feet wide, served this location for the next twenty-nine years. Both bridges were hand operated and made use of the same substructure and timber approaches. In

| Current (fifth) North Avenue Bridge | | | | | |
|---|---|---|---|---|---|
| Opened | Bridge type | Designed by | Constructed by | Cost | Status |
| 2008 | Fixed, cable-stay suspension | Division of Bridges and Viaducts, City of Chicago, and architects Muller+Muller, Ltd. | McHugh Construction | $21.4 million | Currently in use |

Looking north at the current North Avenue Bridge in 2010. © Kevin Keeley

## First, second, and third North Avenue Bridges

| Opened | Bridge type | Designed by | Constructed by | Cost | Status |
|--------|-------------|-------------|----------------|------|--------|
| 1858 | Pontoon swing, wood, hand operated | Unknown | Department of Public Works, City of Chicago | $3,000 | Failed during winter of 1864–65 |
| 1866 | Swing, wood, hand operated | Unknown | Newton Chapin | $3,700 | Removed in 1877 |
| 1877 | Swing, wood and iron, hand operated | Unknown | Conro & Carkin Co. | $7,149 | Removed in 1906 |

1895 this third North Avenue Bridge received a new floor system that carried electric streetcars. Three years later, the height of the bridge was raised, the approaches received extensive repair, the turntable rollers and center pivot were replaced, and new end rollers were installed.

By the early 1900s, the confluence of ship traffic at the north end of Goose Island, combined with the third North Avenue Bridge's center pier, created a serious hindrance to navigation. As a result, the bridge was removed in favor of a new bascule bridge. A temporary bridge opened on January 9, 1906, to maintain traffic. That year, the federal government further improved the river, adding a turning basin just below North Avenue by removing the northwest tip of Goose Island and a portion of the west bank and dredging the river.

The fourth North Avenue Bridge, completed in 1907, was the seventh Chicago-type bascule ever built. The bridge superstructure followed the design of the West Division Street Bridge built in 1904, and its substructure and operating machinery matched the improved layout of the North Western Avenue Bridge the same year. Designed by Thomas G. Pihlfeldt with assistance from Alexander von Babo, the bridge measured 260 feet from end to end, with 173 feet between the trunnions. It had fixed steel girder

approaches and three 115-foot riveted steel trusses to support its symmetrical leaves. A 42-foot-wide roadbed accommodated two lanes of street traffic within the trusses, and sidewalks were cantilevered off of each side.

In 1910 the eastern approach received a new bulkhead, curb wall, and danger-signal system. The stringers under the floor breaks were shimmed up, the roadway planking was patched, a new telephone cable was laid to the bridge houses, and the bridge machinery was overhauled. The following year, the roadway was paved with sectional wood pavement outside the streetcar tracks, the chord covering was patched, the center locks were repaired, the submarine danger-signal cable was repaired and relaid, and the bridge houses were painted. In 1937 the 4-inch block paving was removed and replaced with 3-by-6-inch planking overlaid by 1½ inches of mineral-surface asphalt planking; in addition, the sidewalks, supporting joists, and chord covering were replaced. In 1955 one of the counterweights dropped into the counterweight pit below. Inaccessible from street level, this prompted a $400,000 rehabilitation project to restore the bridge to service.

This bridge, situated in a major industrial corridor, served the City of Chicago for nearly one hundred years. After gentrification

## Fourth North Avenue Bridge

| Opened | Bridge type | Designed by | Constructed by | Cost | Status |
|--------|-------------|-------------|----------------|------|--------|
| Sept. 21, 1907 | Chicago type, double-leaf bascule, steel, electric powered | Division of Bridges and Viaducts, City of Chicago | Roemheld & Gallery (superstructure), Jackson & Corbett (substructure) | $196,964 | Closed to traffic in June 2006 and then removed |

West approach of the fourth North Avenue Bridge looking east in 2006. © Laura Banick

of the neighborhood dramatically increased the traffic pressures around Goose Island and the Lincoln Park neighborhoods in the 1990s, automobile traffic attracted by a boom in retail shopping along the Clybourn Corridor made this two-lane bridge a serious bottleneck. In 2006 the City initiated a $21.4 million project to replace this bridge with a new fixed bridge. They attempted to sell the old bridge, but the relocation costs were prohibitive, and no one stepped up to take the structure. Once a temporary two-lane bridge opened just south of the existing bridge in June 2006, the old steel bridge was dismantled and sold for scrap in the spring of 2007.

# COURTLAND STREET BRIDGES

LOCATION: 1500 West, 1800 North; Courtland Street (formerly Clybourn Place) runs east and west and crosses the North Branch four miles from the river mouth at Lake Michigan.

HISTORICAL HIGHLIGHT: Courtland Street is the location of the first-ever Chicago-type bascule bridge, based on the Tower Bridge in London. Chicagoans initially claimed the first fixed-trunnion bascule in America; it was quickly pointed out, however, that the Grand Avenue trunnion bascule bridge, finished three months earlier in Milwaukee, Wisconsin, designed and patented by Max G. Schinke, was actually the first.

This first Chicago-type fixed-trunnion bascule bridge opened to traffic on May 24, 1902. A prototype for the future of Chicago bridges, it was built under the direction of City engineer John Ericson and incorporated many of the suggestions made by a board of consulting engineers. Other Chicago-type bascules were already under way at Division and 95th streets, but the Courtland Street Bridge was completed first. In 1982 this bridge received landmark designation honors from the American Society of Civil Engineers.

The Courtland Street Bridge looks much as it did when first constructed, with a few notable updates. In 1923 the floor string-ers were replaced, the buffers were replaced, and the bridge was redecked. In 1951 the streetcar rails and old timber deck were removed, and the floor beams and almost all of the stringers were replaced. Several truss members and gusset plates were replaced, and the counterweight boxes were completely rebuilt. Concrete-filled, open steel grating was installed on the back eight feet of each leaf, the sidewalks were replaced, and the approach roadways were rebuilt with concrete slabs. New handrails were installed, and the entire structure was repainted. In 1968 the bridge was again redecked (at a cost of $515,000) with an open-grate steel deck to replace the timber and asphalt floor. In the mid-1990s, the bridge was converted into a fixed span. The operating equipment was decommissioned, and I-beams were bolted across the center of the bridge to lock the two leaves together. In 1997 the bridge superstructure and roadbed were completely restored and repainted by the City of Chicago. It should be noted that an old railroad swing bridge, just south of Courtland Street, is still operational, which allows train cars carrying scrap steel, from A. Finkl & Sons Company and scrap dealers on the east riverbank, to be hauled away.

An interesting element of the Courtland Street Bridge is the raised markings reading "Carnegie" visible on many of the steel beams. Andrew Carnegie, who got started in the steel business

| Current (fifth) Courtland Street Bridge | | | | | |
|---|---|---|---|---|---|
| Opened | Bridge type | Designed by | Constructed by | Cost | Status |
| May 24, 1902 | Chicago type, double-leaf bascule, steel, electric powered | Division of Bridges and Viaducts, City of Chicago | American Bridge Co. (superstructure), FitzSimons & Connell Co. (substructure) | $152,911 | Currently in use |

Looking west at the current Courtland Street Bridge in 2012. © Patrick McBriarty

in the 1870s, later formed the Carnegie Steel Company in 1892, which was sold in 1901 to United States Steel, or U.S. Steel, as it is commonly known. The steel incorporated in this bridge may have been some of the very last Carnegie steel ever produced.

Though the first Courtland Street Bridge, located at what was then called Clybourn Place, was not part of Chicago, it was named after a prominent early Chicagoan, butcher, and the town's first constable, Archibald Clybourne. Clybourne moved to Chicago in 1826 and is credited with launching Chicago as a major center for livestock and slaughterhouses. In 1828 he secured an important government contract to supply meat for Fort Dearborn, Fort Mackinac, Fort Howard, and Fort Winnebago. His businesses having done quite well, in 1836 he moved his family into a twenty-room house on Elston Road (now Elston Avenue) near Clybourn Place, just west of the river. His residence is noted on the 1849 map shown here. The only known documentation of a bridge at Clybourn Place is this map and an 1859 Common Council order for its replacement by a drawbridge.

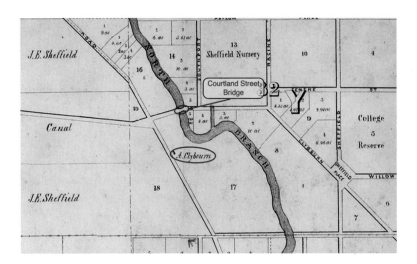

Map showing the first Courtland Street Bridge and Archibald Clybourne's residence in 1836.

The second Courtland Street Bridge was built in 1859. Residents and business owners demanded a moveable bridge, as the dredging and widening of the North Branch for shipping had recently reached Clybourn Place and beyond. The river separated

| First and second Courtland Street Bridges | | | | | |
|---|---|---|---|---|---|
| Opened | Bridge type | Designed by | Constructed by | Cost | Status |
| pre-1849 | Fixed, wood bent | Unknown | Unknown | Unknown | Removed in 1859 |
| Sept. 1859 | Fixed, wood bent | Unknown | Unknown | Unknown | Removed on Apr. 28, 1862 |

the towns of West Chicago and City of Chicago, so construction of this new bridge required cooperation between the county and city governments. The resulting bridge was a fixed bridge, prompting petitions just a year later to replace it with a moveable bridge because it was a hindrance to navigation.

Three years later, its attempted removal resulted in civil unrest that culminated in a bloodless battle. In attempting to tear down the old structure, the contractor and his workers were chased off by West Chicagoans several times. The West Side commuters vowed to protect the old bridge until the contractor was prepared with the materials to promptly erect the new bridge. The situation came to a head on Monday, April 28, 1862, when a party of twenty-five police officers landed by tugboat to confront forty to fifty residents led by Archibald Clybourne. The police, "armed to the teeth" with bludgeons and hickory clubs, faced off against vigilantes armed with spades, shovels, picks, and various other weapons.[1] With "death in the eyes" of the crowd and "fearless resolve" on the part of the constables, the contending parties were about to clash when a flag of truce on a pitchfork handle was observed.[2] A parley ensued, and the contractor "negotiated satisfactorily," allowing all parties to return home unharmed and with egos intact.[3] The second Courtland Street Bridge was torn down later that day following assurances that the materials were indeed on hand and that construction of a new bridge would proceed swiftly.

The third Courtland Street Bridge was praised as "a neat, handsome little structure" of 100 feet in length that worked "beautifully" by rotating on its bearings and thus allowing one operator to turn it quickly and without difficulty.[4] This drawbridge was replaced in 1873.

| Third and fourth Courtland Street Bridges | | | | | |
|---|---|---|---|---|---|
| Opened | Bridge type | Designed by | Constructed by | Cost | Status |
| 1862 | Pivot, wood, hand operated | Fox & Howard | Fox & Howard | $2,000 | Removed in 1873 |
| 1873 | Swing, wood, hand operated | Fox & Howard | Fox & Howard | $13,700 | Removed between Dec. 10, 1900, and Jan. 1, 1901 |

Looking southeast at the fourth Courtland Avenue Street Bridge in the open position over the protection pier in 1900.

The fourth Courtland Street Bridge was a second, larger, Fox & Howard swing bridge, 140 feet long by 32 feet wide. The *Chicago Tribune* praised the firm for "unsurpassed facilities" in dredging and bridge building and recommended that corporations and individuals consult them.[5]

In early December 1900, the FitzSimons and Connell Company, under a City contract, began removal of the twenty-three-year-old bridge. The superstructure, center pier, and abutments were removed from the river. The project was completed on January 1, 1901, making way for the very first Chicago-type bascule bridge.

# WEBSTER AVENUE BRIDGES

LOCATION: 1600 West, 2200 North; Webster Avenue runs east and west, crossing the North Branch 4.9 miles from the river mouth at Lake Michigan.

Today's Webster Avenue Bridge is a Chicago-type bascule bridge designed by chief bridge engineer Thomas Pihlfeldt and City bridge engineer Alexander von Babo. Completed in 1916, it was the seventh of twenty-eight second-generation Chicago-type bascules built between 1913 and 1938. This bridge presents a 161-foot clear river channel between the masonry approaches and measures 189 feet between the trunnions. The only river crossing between Courtland Street and Fullerton Avenue, this became a key North Side connection (as Ashland Avenue did not receive a bridge until twenty years later).

In 1936 the 4-inch creosoted wood-block paving was removed and replaced with 3-by-6-inch deck overlaid by 1½ inches of asphalt planking. In addition, new cast-steel floor breaks replaced the timber framing. The bridge was reopened in conjunction with the opening of the Ashland Avenue Bridge. In 1968 the Webster Avenue Bridge received a new open-grid steel deck and general repairs. In the mid-1990s, it was converted to a fixed span; as in the case of other such bridges, two large I-beams were incorporated to lock the two bridge leaves together. The northwestern bridge house was removed, and the southeastern bridge house, its top half already missing, was secured and abandoned. There are no identifying plaques on the bridge; the only marking, a stenciled "5-06" on the superstructure, denotes the bridge's last painting to date in May 2006.

The first bridge at Webster Avenue used the refurbished Clark Street Howe-truss superstructure, originally built in 1872. Work to move the bridge began at eight o'clock on the night of Saturday, April 13, 1889. A temporary framework was constructed on scows under each half of the bridge superstructure. Water ballast was pumped out of the scows to lift the bridge and float it off the foundations. In preparation, the scows had been flooded with enough water to more than equal the calculated weight of the structure. Once the bridge was raised, two tugs, one in front and one behind, towed it five miles upriver to Webster Avenue. The trip was completed in just five hours without incident. The experience gained from moving the Wells Street Bridge to Dearborn Street the year before made this undertaking straightforward.

With the substructure still under construction, the bridge was placed ashore on the eastern side of the street, where it was

| Current (second) Webster Avenue Bridge | | | | | |
|---|---|---|---|---|---|
| Opened | Bridge type | Designed by | Constructed by | Cost | Status |
| Aug. 3, 1916 | Chicago type, double-leaf bascule, steel, electric powered | Division of Bridges and Viaducts, City of Chicago | Ketler-Elliott Co. (superstructure), Great Lakes Dredge & Dock Co. (substructure) | $285,558 | Currently in use |

Looking south at the current Webster Avenue Bridge in 2007. © Laura Banick

## First Webster Avenue Bridge

| Opened | Bridge type | Designed by | Constructed by | Cost | Status |
|---|---|---|---|---|---|
| 1889 | Swing, iron, hand operated | Fox & Howard | Fox & Howard (reused superstructure), Chicago Dredge & Dock Co. (substructure) | $11,500 | Removed on June 4, 1914 |

thoroughly overhauled, repaired, and repainted. Once the new foundation was ready, the City Bridge Department installed the bridge on the new substructure, and the first Webster Avenue Bridge opened, almost as good as a new bridge. After twenty-five years of service, this bridge was removed by the Great Lakes Dredge & Dock Company at a cost of $4,127, in June 1914.

The first Webster Avenue Bridge in 1911.

# NORTH ASHLAND AVENUE BRIDGE

LOCATION: 1600 West, 2201 North; Ashland Avenue runs north and south, crossing the North Branch 4.9 miles from the river mouth at Lake Michigan.

The current North Ashland Avenue Bridge is the first at this location. The opening of this bridge in 1936 completed the last link in a $27.5 million Ashland Avenue improvement from 95th Street to Devon Avenue. Started in 1922, the project made Ashland Avenue a major north-south boulevard, as recommended in the *Plan of Chicago*. This second-generation Chicago-type bridge carrying four lanes of traffic was paid for with a PWA grant of $485,500 from the City's share of state gasoline-tax funds. At the ribbon-cutting ceremony, Mayor Edward J. Kelly declared that the entire city would benefit from the Ashland improvements, which were "a concrete expression of the 'I WILL' spirit of Chicago."[1]

In 1993 this bridge underwent complete refurbishment. In 1995 the Army Corps of Engineers agreed that, since most bridges above North Halsted Street had not been opened for several years, they effectively functioned as fixed bridges. This relieved the City of the responsibility and expense of maintaining their mechanical systems. The bridge was converted to a fixed span. Several years later, the balance of the City-owned North Branch bridges were also converted to fixed spans.

| North Ashland Avenue Bridge | | | | | |
|---|---|---|---|---|---|
| Opened | Bridge type | Designed by | Constructed by | Cost | Status |
| Aug. 20, 1936 | Chicago type, double-leaf bascule, steel, electric powered | Division of Bridges and Viaducts, City of Chicago | Ketler-Elliott Co. (superstructure), FitzSimons & Connell Co. (substructure) | $1.7 million | Currently in use |

Looking north at the North Ashland Avenue Bridge in 2007. © Laura Banick

# FULLERTON AVENUE BRIDGES

LOCATION: 2400 North, 1850 West; Fullerton Avenue runs east and west, crossing the North Branch 4.8 miles from the river mouth at Lake Michigan.

HISTORICAL HIGHLIGHT: The fourth Fullerton Avenue Bridge was completed in 1961 and was the first modern fixed bridge placed over the navigable portion of the North Branch since federal oversight of the river began in 1890.

This fixed-steel and reinforced-concrete span is 277 feet long and 80 feet wide and was built in two stages. The two northern lanes were built first, while the old swing bridge continued to carry traffic, from June to December 1960. Then traffic was routed to the new portion of the bridge, the old swing bridge was removed, and the remaining four lanes of the bridge were built. The completed six-lane bridge opened on September 21, 1961, offering a 21-foot river clearance. Even with its more expensive high-grade approaches for river clearance, the bridge saved city taxpayers an estimated $825,000 in construction costs and $29,000 per year in bridge-tender and maintenance expenses when compared with a moveable bridge.

By the 1960s, the City had been working toward a fixed-bridge policy on the North Branch for decades. Earlier federal oversight regulations required drawbridges to offer a minimum 16½-foot river clearance while in the closed position. After gaining federal approval for a 60-foot fixed bridge for the Dan Ryan Expressway, the City also finally received the necessary approvals to build the Fullerton Avenue Bridge. This was the first concrete step in developing the City's long-sought fixed-bridge policy for the North Branch. The *Public Works Annual Report of 1960* predicted that this would be the first of eight new fixed bridges over the North Branch; so far, seven fixed bridges have replaced drawbridges. The other six on the North Branch are at North, Diversey, Belmont, North Damen, and North Western Avenues and North Halsted Street over the North Branch Canal.

Before the first Fullerton Avenue Bridge in 1874, the street began at the lakefront and ended at Ashland Avenue. Charles Fullerton owned the tract of land west of Ashland Avenue and donated land to extend the roadway west for the construction of the first Fullerton Avenue Bridge. Fullerton Avenue was the dividing line between Chicago and the town of Lake View to the north. The two municipalities shared in the cost of this first bridge. Lakeview provided one-quarter, and the City of Chicago provided three-quarters of the necessary funds. The result was a 225' × 20' fixed iron bridge. The bridge was removed in 1877 to make way for a drawbridge and allow for shipping farther up the North Branch.

The second Fullerton Avenue Bridge was a wood and iron swing bridge built under a similar scheme. One-third of the funds

| | Current (fourth) Fullerton Avenue Bridge | | | | |
| --- | --- | --- | --- | --- | --- |
| Opened | Bridge type | Designed by | Constructed by | Cost | Status |
| Sept. 21, 1961 | Fixed, steel and concrete | Division of Bridges and Viaducts, City of Chicago | William E. Sweitzer & Co. | $1.35 million | Currently in use |

Looking north at the current Fullerton Avenue Bridge in 2012. © Patrick McBriarty

## First and second Fullerton Avenue Bridges

| Opened | Bridge type | Designed by | Constructed by | Cost | Status |
|---|---|---|---|---|---|
| 1874 | Fixed, iron | Fox & Howard | Fox & Howard | $5,000 | Removed in 1877 |
| 1877 | Swing, wood and iron, hand operated | Unknown | I. W. Lavin Co. | $7,444 | Removed in 1894 or 1895 |

came from the Town of Lake View and two-thirds of which came from the City of Chicago. This new swing bridge was the same size as the previous fixed bridge.

Founded in 1857, the town of Lake View was bound on the south by Fullerton Avenue, on the west by Western Avenue, on the north by Devon Avenue, and on the east by Lake Michigan. The town hall, built in 1872 at the corner of Halsted and Addison streets, is the site of the old 11th District Police Station, which has been mostly replaced by a new larger facility a half block west. The town of Lake View became a city in 1887, and in 1889 the Lake View City Council, in a controversial vote, allowed its annexation by the City of Chicago. Thereafter, Chicago bore sole responsibility for the Fullerton Avenue Bridge. By 1895 the poor condition of this bridge led to its removal.

The third Fullerton Avenue Bridge, completed in 1895, was one of the last swing bridges ever built over the Chicago River. It was a Pratt thru-truss superstructure designed to also carry the streetcars of the North Chicago Street Railroad Company, which contributed eighteen thousand dollars toward the bridge's construction. The two trusses were 22 feet from center to center, offering an 18-foot roadway with two 5-foot sidewalks on either side. The overall length was 164½ feet. The bridge rested on a rim-bearing turntable, originally operated by hand and converted to electric power by the streetcar company. It was first operated under electric power on January 1, 1899. The time it took to open or close the bridge was reduced to thirty-five to forty seconds.

In 1917 the U.S. Army Corps of Engineers declared this bridge an obstruction to navigation. The bridge was not removed,

## Third Fullerton Avenue Bridge

| Opened | Bridge type | Designed by | Constructed by | Cost | Status |
|---|---|---|---|---|---|
| 1895 | Swing, steel, electric powered | Chicago Bridge & Iron | Chicago Bridge & Iron (superstructure), FitzSimons & Connell Co. (substructure) | $27,753 | Removed in Dec. 1960 |

Aerial view looking northeast of the third Fullerton Avenue Bridge in the 1950s. Courtesy of CDOT.

however, and, beginning in 1938, the City of Chicago made repeated applications to replace the bridge with a fixed span. Despite the significant decline in river traffic on the North Branch, the City's proposals were firmly rebuffed by federal authorities. In 1945 the City submitted plans to federal authorities for a short, inexpensive moveable bridge, which were accepted but never built. In 1955 the U.S. Army Corps of Engineers agreed to support City efforts for a fixed bridge at Fullerton. The old swing bridge was kept in place until December 1960, when it was removed for construction of the remaining portion of the new fixed bridge.

# NORTH DAMEN AVENUE BRIDGES

LOCATION: 2000 West, 2560 North; Damen Street (formerly Robey Street) runs north and south, crossing the North Branch 5.6 miles from the river mouth at Lake Michigan.

HISTORICAL HIGHLIGHT: Today's North Damen Avenue Bridge was the first fixed-arch suspension bridge built in the United States.

The new Damen Avenue Bridge, which opened in 1999, supports its suspended bridge deck with striking twin arches. This unique bridge replaced a sixty-nine-year-old single-leaf Chicago-type bascule. The revolutionary design was a first for Chicago when built. In 2005 traffic was reduced from four to two lanes so that bike lanes could be added on each side of the roadway. In July 2011, in much need of some attention, the bridge received a fresh coat of light-gray paint to cover its former iconic Chicago Bulls–red paint job.

The Chicago architectural firm of Muller+Muller Architects, designed this bridge, the new North Avenue suspension bridge, and the new North Halsted Street tied-arch fixed bridge over the canal. The bridge engineer for the North Damen Avenue Bridge was J. Muller International (no relation), and the Transystems Corporation provided civil engineering support. Eighty percent of the funding for this bridge came from the federal government, and the balance came from State of Illinois coffers. The bridge's single-piece steel tubes, each weighing seventy-four tons, were manufactured in Duluth, Minnesota, and shipped by barge from Lake Superior to Lake Michigan and then up the Chicago River.

The first North Damen Avenue Bridge was a 162-foot single-leaf Chicago-type bascule built by the Board of Local Improvements and opened in 1929. The bridge utilized a Pratt-truss superstructure, and the octagonal bridge houses were built in the Beaux Arts style with a band of windows, decorative frieze above and below, and hipped roofs topped by a decorative vent. The bridge provided a 44-foot-wide roadway and 9-foot-wide sidewalks and featured two ramped approaches that connected it to the street. The 1,000-foot-long southern approach was backfilled, whereas the northern approach included a 1,140-foot steel and concrete viaduct and a 280-foot approach ramp. The greater elevation and viaduct gave the International Harvester Company a connection between its yards on either side of the formerly dead-end street along the river. The Public Works Administration purchased the land from International Harvester in 1935 as part of a plan to develop public housing on the site. Completed in 1938, the Julia C. Lathrop Homes, totaling 925 units on 35.5 acres, constituted one of Chicago's first public

| Current (second) North Damen Avenue Bridge | | | | | |
|---|---|---|---|---|---|
| Opened | Bridge type | Designed by | Constructed by | Cost | Status |
| 1999 | Fixed, arch suspension | TranSystems, J. Muller International, Muller+Muller, Ltd. | Walsh Construction Co. | $12.8 million | Currently in use |

Looking north at the current North Damen Avenue Bridge in 2010. © Kevin Keeley

housing projects and within the past decade have been recommended for historic preservation.

Eugene S. Taylor, manager of the Chicago Plan Commission, claimed that a continuous Damen Avenue from 74th Street north to Bryn Mawr Avenue would benefit one million people—roughly a third of Chicago's citizens at that time. The bridge's opening was celebrated by thousands of residents from the Northwest Side as "the greatest improvement that section of the city had seen."[1] Prior to this bridge being built, only Webster Avenue and Western Avenue, located two miles apart from one another, offered North Branch crossings in the area.

In January 1998, the street was closed, and this bascule bridge was razed for replacement by a new fixed bridge.

Looking north at the first North Damen Avenue Bridge in the 1950s. Courtesy of CDOT.

## First North Damen Avenue Bridge

| Opened | Bridge type | Designed by | Constructed by | Cost | Status |
|---|---|---|---|---|---|
| Jan. 24, 1929 | Chicago type, single-leaf bascule, steel, electric powered | Division of Bridges and Viaducts, City of Chicago | Great Lakes Dredge & Dock Co. | $1.65 million | Removed in Jan. 1998 |

# DIVERSEY AVENUE BRIDGES

LOCATION: 2150 West, 2800 North; Diversey Avenue runs east and west, crossing the North Branch six miles from the river mouth at Lake Michigan.

Diversey Avenue received the second fixed bridge over the navigable section of the North Branch. Following the new Fullerton Avenue fixed bridge built in 1961, Diversey's new five-lane steel and concrete bridge, with a 22-foot river clearance, greatly improved the connection between the North Side and West Side boulevard systems. It replaced the last City-owned swing bridge on the Chicago River. However, two railroad swing bridges still stand on the North Branch: one has been converted to also allow bikes and pedestrians to connect with Goose Island just below North Avenue; the other is an operational plate-steel swing bridge that is usually open near Finkle Steel, just south of Courtland Street.

The first Diversey Avenue Bridge was 184 feet long by 35 feet wide—20 feet longer than, but otherwise very similar to, the 1895 Fullerton Avenue swing bridge. Both bridges were Pratt thru-truss swing bridges and were constructed a year apart. The Diversey Bridge was initially hand operated and had a concrete-capped pile center pier, with pile and timber protection, pile abutments, and trestle bent approaches.

On October 20, 1900, police captain "Barney" Baer received an involuntary swim at this location. The bridge was open while he was patrolling Diversey Boulevard west of the river when a noise startled his horse. With little or no warning, the animal bolted, carrying the captain and his buggy over the edge and plunging into the river below. Clinging to the reins, the captain stayed with the horse as it gradually swam toward the shore, slowed by the still-attached buggy. Man and beast were pulled to safety, and the buggy was recovered and hauled to a blacksmith for repair. Meanwhile, the apparently embarrassed Baer attempted to suppress the accident by drying his clothes in the engine room of a nearby factory; his aquatic adventures did not go unnoticed, however, and were reported in the *Chicago Daily Tribune* the next day.

In 1909 the bridge house was moved from the northeast corner to the southwest corner of the bridge. Subsequently destroyed by a fire in 1921, a new, more attractive bridge house was built on the northwest corner, at a cost of $4,182.41. Between September 17 and September 29, 1929, the entire roadway was removed. The six-by-twelve-inch creosoted pine subplanking

| Current (second) Diversey Avenue Bridge | | | | | |
| --- | --- | --- | --- | --- | --- |
| Opened | Bridge type | Designed by | Constructed by | Cost | Status |
| 1968 | Fixed, steel and concrete | Division of Bridges and Viaducts, City of Chicago | J. M. Corbett Co. | $1,654,712 | Currently in use |

Looking northwest at the current Diversey Avenue Bridge in 2007. © Laura Banick

was reused, but a new three-by-six-inch redwood deck covered by a one-and-a-half-inch asphalt planking was installed. The bridge roadway was redecked in a similar fashion in 1947.

On the afternoon of April 2, 1928, the steamer *Gilbert* collided with the bridge, knocking it off of its turntable and shifting it about eight feet to the southwest. The steel drum of the turntable was badly twisted, and the bridge listed at a precarious angle. Repairs on the bridge were conducted day and night by City workers with assistance from the FitzSimons and Connell Company. The span was jacked up, the turntable was repaired, and the bridge, once returned to its proper position, reopened to traffic on April 12.

Repeated efforts to replace this bridge began in 1935, but this bridge survived for several decades. In 1963, lacking the funds needed for a new span, the then oldest City-owned bridge underwent a $150,000 three-month repair. This work was predicted to keep the bridge "in good shape for at least 10 years."[1] However, in 1967 funds were secured, and it was finally removed to make way for the construction of a modern fixed bridge.

Looking southwest at the first Diversey Avenue Bridge in 1944. Courtesy of CDOT.

## First Diversey Avenue Bridge

| Opened | Bridge type | Designed by | Constructed by | Cost | Status |
|---|---|---|---|---|---|
| Jan. 11, 1896 | Swing, steel, hand operated | Unknown | Lassing Bridge & Iron Co. (superstructure), Lydon & Drews Co. (substructure) | $31,345 | Removed in Feb. 1967 |

# NORTH WESTERN AVENUE BRIDGES

LOCATION: 2400 West, 3000 North; Western Avenue runs north and south, crossing the North Branch 6.4 miles from the river mouth at Lake Michigan.

HISTORICAL HIGHLIGHT: Completion of the first North Western Avenue Bridge in 1891 connected the thoroughfare from Blue Island through Evanston, a distance of more than 30 miles, making it the longest continuous street in Chicago and, possibly, the world.

The City proposed a new fixed bridge at North Western Avenue and received U.S. Coast Guard approval in 1967. Federal approval was bolstered by the fact that the old two-lane bridge had not been opened for two years. A contract for a fixed bridge was awarded at the end of 1971, and construction was completed in 1973. The new four-lane bridge greatly improved automobile traffic in the area. The bridge's 16' 9" clearance over the river was deemed sufficient for any barge or recreational boat traffic using the boatyards north of Western Avenue (i.e. Grebe's Shipyard).

The first North Western Avenue Bridge was a Howe-truss swing proposed in early 1886 and financed by Lake View, Jefferson Township, the City of Chicago, and Cook County. Five years later, the river was widened, the course of the river was straightened, and this 266-foot-long bridge was constructed. The bridge, which opened in 1891, was of the same type as the first swing bridge at Canal Street. It had a single 18-foot-wide roadway that allowed north- and southbound wagons to cross at the same time and 5-foot-wide sidewalks on each side. The wood and iron truss rested on an iron and steel turntable supported by a pile center pier. Pile and timber approaches connected the bridge with the street. The cost of this bridge was paid for entirely by the City of Chicago, but included appropriations earmarked from Jefferson and Lake View (the latter of which had been annexed by Chicago in 1889). After twelve years, the FitzSimons and Connell Company received a contract to remove the bridge in favor of a new bascule bridge. Removal began on January 16, 1903, and was completed by February 1903.

The second North Western Avenue Bridge was similar to the old swing bridge it replaced, of equal width, providing two lanes of traffic and 8-foot sidewalks on either side. This first-generation Chicago-type bascule was 205½ feet between the trunnions and, when open, offered a 168-foot river channel. It provided a much wider channel and no center pier, both significant improvements to navigation. This three-truss, double-leaf bascule bridge constructed from 1,090 tons of structural steel received a 700-ton counterweight on each leaf. The bridge was operated using two 38-horsepower electric motors for each leaf. It took ten months

| Current (third) North Western Avenue Bridge | | | | | |
|---|---|---|---|---|---|
| Opened | Bridge type | Designed by | Constructed by | Cost | Status |
| 1973 | Fixed, steel and concrete | Division of Bridges and Viaducts, City of Chicago | Schless Construction Co. | $2,787,655 | Currently in use |

Looking north at the current North Western Avenue Bridge in 2007. © Laura Banick

to build the bridge foundations and another ten months to construct the superstructure.

In 1911 the bridge was given sectional wood pavement outside the streetcar tracks on either side of the center truss, and the approaches received yellow-pine block pavers. The Civil Works Administration Project (No. 999), completed on March 28, 1934, replaced the steel plates on the rail stringers, repaired the castings at the road breaks, and redecked this bridge. In 1944 the bridge deck and floor breaks were again replaced, and approximately five tons of counterweights were added to each leaf. A failed abutment wall on the northwest corner of the bridge was also rebuilt.

After almost seventy years, this old swing bridge was removed in 1973 and replaced by a four-lane fixed bridge.

The first North Western Avenue Bridge resting over the protection pier in 1903.

Looking southeast at the second North Western Avenue Bridge in 1904.

## First and second North Western Avenue Bridges

| Opened | Bridge type | Designed by | Constructed by | Cost | Status |
|---|---|---|---|---|---|
| 1891 | Swing, Howe truss, hand operated | Unknown | Binder & Seifert (superstructure), Chicago Dredge & Dock Co. (substructure) | $32,706 | Removed in Jan. 1903 |
| Nov. 22, 1904 | Chicago type, double-leaf bascule, steel, electric powered | Division of Bridges and Viaducts, City of Chicago | C. L. Strobel (superstructure), FitzSimons & Connell Co. (substructure) | $303,998 | Removed in 1973 |

# BELMONT AVENUE BRIDGES

LOCATION: 2600 West, 3200 North; Belmont Avenue runs east and west, crossing the North Branch 6.8 miles from the river mouth at Lake Michigan.

HISTORICAL HIGHLIGHT: As the northernmost limit to navigation on the North Branch, Belmont Avenue and all crossings south required moveable bridges from 1890 until the early 1960s.

The fourth and current Belmont Avenue Bridge was also the fourth North Branch drawbridge to be replaced by a fixed span. The four-lane fixed bridge providing a twenty-one-foot clearance replaced a two-lane Chicago-type bascule bridge in 1976.

During the summer of 2010, a three-foot alligator was spotted under this bridge and then captured by local wildlife conservationist Bob the Alligator Wrangler. (Bob, who does not like to give out his last name, is a volunteer for the Chicago Herpetological Society.) The alligator was held for several months while arrangements were made for its transfer to a Florida group that could see to its release back into the wild. It was not the first time in recent history that such a creature was caught in the Chicago River; a four-and-a-half-foot, forty-five-pound alligator was captured in June 2008.

In 1875, when the first Belmont Avenue Bridge was built, the northern boundary of Chicago was Fullerton Avenue. A mile north, Belmont Avenue connected Lake View east of the river with Jefferson to the west. The two municipalities and Cook County shared in paying for the first Belmont Avenue Bridge. Of the $3,290 total cost, Jefferson paid $1,491, Lake View paid $1,097, and Cook County paid $694. In 1889 the City of Chicago annexed Lake View and Jefferson and took ownership of this seventy-seven-foot-long by nineteen-foot-wide bridge. The City paid to refurbish it the same year.

The second Belmont Avenue Bridge utilized the Canal Street Bridge, which was ordered removed in June 1892 and was subsequently floated to Belmont Avenue and placed on the east bank of the river. Plans were drawn, bids opened, and construction begun on a foundation for a moveable bridge on October 27. The work was halted three days later by neighboring landowners, however. An injunction by F. W. H. Sandmacher and H. J. C. Glade prevented the City or the contractor from encroaching on their land. Since this property abutted the approach to the

| Current (fourth) Belmont Avenue Bridge | | | | | |
|---|---|---|---|---|---|
| Opened | Bridge type | Designed by | Constructed by | Cost | Status |
| 1976 | Fixed, steel and concrete | Division of Bridges and Viaducts, City of Chicago | Unknown | Unknown | Currently in use |

Looking east at the current Belmont Avenue Bridge in 2008. © Laura Banick

Belmont Avenue Bridge, it effectively halted removal of the Belmont Avenue Bridge. On May 13, 1893, the City purchased the land for the sum of $5,500, dissolving the injunction and allowing work to resume on June 10. The old fixed bridge was removed, and the new foundation was completed by August 31, 1893.

In October 1893, the two-year-old swing bridge from Canal Street was installed on the new foundation to become the second Belmont Avenue Bridge. The Howe thru-truss bridge was two hundred feet long and thirty-five feet wide and accommodated the now wider and deeper North Branch, which had been dredged prior to installation of the new bridge.

On July 10, 1910, Riverview Amusement Park employees searched the river from Lawrence to Belmont Avenue for an escaped boa constrictor. Thousands watched from the shores and bridges, wondering who had drowned, as workers dredged the river. At the height of the search, the big snake wriggled onto the riverbank near Belmont Avenue, causing panic among some children playing in the vicinity. After a short struggle, it was recaptured and taken back to its cage.

Soon enough a third bridge was necessary to replace the Belmont swing bridge with a new bascule bridge. Due to labor problems and the refusal by neighboring landowners to allow a temporary bridge, however, it became a complicated six-year process. In order to maintain traffic at this important crossing, a series of three temporary bridges were constructed. The first, which opened on June 16, 1911, was a temporary pedestrian bridge placed just north of the old structure within the street lines of Belmont Avenue. The following day, work began to remove the old swing bridge. The superstructure was floated on scows and towed to the turning basin below North Avenue. There, it was refitted while the old foundation and center pier were removed. A temporary pivot pier and abutment were constructed, and the refitted bridge was moved from the turning basin and installed as a second temporary bridge. This bridge opened to streetcar, horse-drawn, and automotive vehicles on August 5, 1911, and the temporary pedestrian bridge was removed the next day.

In 1912 plans were prepared for a third temporary bridge that would both carry traffic and allow construction of the new

### First and second Belmont Avenue Bridges

| Opened | Bridge type | Designed by | Constructed by | Cost | Status |
|--------|-------------|-------------|----------------|------|--------|
| 1875 | Fixed, iron | King Bridge Co. | King Bridge Co. | $3,290 | Removed in 1893 |
| Oct. 4, 1893 | Swing, Howe truss, hand operated | Abraham Gottlieb, civil engineer | A. Gottlieb & Co. (superstructure), Chicago Dredge & Dock Co. (substructure) | $24,759 | Moved to the turning basin near North Avenue on June 17, 1911 |

Looking southeast at a warship passing through the third Belmont Avenue Bridge in 1943. Courtesy of the Chicago Maritime Museum.

## Third Belmont Avenue Bridge

| Opened | Bridge type | Designed by | Constructed by | Cost | Status |
|---|---|---|---|---|---|
| July 12, 1917 | Chicago type, double-leaf bascule, steel, electric powered | Division of Bridges and Viaducts, City of Chicago | Great Lakes Dredge & Dock Co. | $268,033 | Removed in 1976 |

Panorama of Grebe's Yacht Yard with warships under construction in 1942. Courtesy of the Chicago Maritime Museum.

bascule bridge in the upright position without encroaching upon the neighboring properties. This scow bridge was constructed of steel and designed so that it could be used repeatedly at other localities. The second temporary bridge was removed by the end of 1912, and the third temporary bridge was floated into place. It opened to traffic on January 3, 1913; it was finally closed to traffic on October 17, 1916, and was removed on November 9.

Construction of the third Belmont Avenue Bridge began in 1912. Two labor strikes in 1915 halted construction, causing most of the delay that year. The leaves of the new bascule were lowered for the first time on May 10, 1916. Installation of streetcar rails, pavement, and sidewalks followed, and the bridge was opened to pedestrians and a single streetcar on October 17, 1916. Work on the superstructure, nearly identical to the Webster Avenue Bridge built in 1916, was finally completed on February 6, 1917. Additional projects, such as bricking over the pavement

and curbs on the approaches, were completed on June 24, and the bridge was officially opened on July 12, 1917.

This bridge had two notable neighbors that, like this bridge, no longer exist. To the northeast was the Riverview Amusement Park, which was open from 1903 to 1967. To the northwest was the Henry C. Grebe & Company shipyard, open from 1926 to 1993.[1] Grebe produced a variety of pleasure yachts and warships, most notably the 136-foot Mark I and II minesweepers used during World War II and the Korean War. The last Grebe minesweeper passed down the North Branch and was delivered in 1953. As an important U.S. Navy contractor, the existence of Grebe at this location likely thwarted City efforts to adopt a fixed-bridge policy on the North Branch until the 1960s.

In 1976 the third Belmont Street Bridge was replaced by a new fixed four-lane bridge.

**Fixed Wood Bent Bridges**

| Location | No. | Built | Removed | Waterway | Notes |
|---|---|---|---|---|---|
| Kinzie Street | 1st | 1832 | 1839 | North Branch | Built by Samuel Miller |
| South Branch Bridge | 1st | 1832 | 1838 | South Branch | Built by Charles & Anson Taylor via private subscriptions |
| Pearson Street* | 1st | Pre-1836 | Unknown | North Branch | Built by private concerns |
| Union Street* | 1st | Pre-1836 | Unknown | North Branch | Built by private concerns |
| Unnamed crossing (possibly Hickory St.) | 1st | Pre-1836 | Unknown | South Fork of S.B. | Built by private concerns; existed before Archer Avenue just north of street line |
| Archer Avenue | 1st | Pre-1836 | Unknown | Healey's Slough | Built by private concerns, just west of Green St. |
| Archer Avenue | 1st | Pre-1836 | 1848 | South Fork of S.B. | Built by private subscriptions* |
| Archer Avenue | 2nd | 1848 | Unknown | South Fork of S.B. | Built by private subscriptions* |
| Fuller Street (formerly Bridge Street) | 1st | Pre-1849 | 1865 | South Branch | Built by private subscriptions* |
| S. Ashland Avenue (W.F.) | 1st | Pre-1849 | Unknown | West Fork of S.B. | Built by private subscriptions* |
| S. Ashland Avenue (I&M Canal) | 1st | Pre-1849 | 1849 | I&M Canal | Built by private subscriptions* |
| Courtland Avenue (formerly Clybourn Place) | 1st | Pre-1849 | 1859 | North Branch | Built by private subscriptions* |
| S. Ashland Avenue (I&M Canal) | 2nd | 1849 | Unknown | I&M Canal | |
| Courtland Avenue | 2nd | 1859 | 1862 | North Branch | |
| S. Ashland Avenue (I&M Culvert) | 1st | 1881 | 1908 | Culvert of I&M Canal | |
| S. Ashland Avenue (I&M Culvert) | 2nd | 1908 | 1930s | Culvert of I&M Canal | |

*Note:* S.B. = South Branch; S.F. = South Fork of the South Branch; W.F. = West Fork of the South Branch

*Believed to be financed by private subscription, but clear documentation was not found.

# DESIGN TYPE

**Fixed Iron Bridges**

| Location | No. | Built | Removed | Waterway | Notes |
|---|---|---|---|---|---|
| Fullerton Avenue | 1st | 1874 | 1877 | North Branch | Iron truss |
| Belmont Avenue | 1st | 1875 | 1893 | North Branch | Iron truss |
| S. Ashland Avenue (I&M Canal) | 3rd | 1886 | 1909 | I&M Canal | Iron truss |

**Fixed Steel Bridges**

| Location | No. | Built | Removed | Waterway | Notes |
|---|---|---|---|---|---|
| S. Ashland Avenue (I&M Canal) | 4th | 1909 | Unknown | I&M Canal | Removed sometime after 1933 |

**Fixed Steel and Concrete Bridges**

| Location | No. | Built | Removed | Waterway | Notes |
|---|---|---|---|---|---|
| Fullerton Avenue | 4th | 1961 | Present | North Branch | 21' river clearance |
| Dan Ryan Expressway (I-90/I-94) | 1st | 1965 | Present | South Branch | 60' river clearance |
| Diversey Avenue | 2nd | 1968 | Present | North Branch | 21' river clearance |
| N. Western Avenue | 3rd | 1973 | Present | North Branch | 21' river clearance |
| Belmont Avenue | 4th | 1976 | Present | North Branch | 21' river clearance |
| Lake Shore Drive, Ogden Canal | 2nd | 1986 | Present | Michigan Canal | Cluster of nine fixed bridges on two levels |
| Damen Avenue | 2nd | 1999 | Present | North Branch | Steel tube fixed-arch suspension bridge |
| North Avenue | 5th | 2008 | Present | North Branch | Cable-stayed suspension bridge |
| N. Halsted Street (canal) | 3rd | 2012 | Present | North Branch | Fixed-arch suspension bridge |

## Wood Gallows-Frame Drawbridge

| Location | No. | Built | Removed | Waterway | Notes |
|---|---|---|---|---|---|
| Dearborn Street | 1st | 1834 | 1839 | Main Channel | First drawbridge in Chicago; two leaves, no counterweights |

## Pontoon Swing Bridges

| Location | No. | Built | Removed | Waterway | Notes |
|---|---|---|---|---|---|
| Randolph Street | 1st | 1838 | 1847 | South Branch | Built by private subscriptions* |
| Kinzie Street | 2nd | 1839 | 1849 flood | North Branch | Built by private subscriptions* |
| Clark Street | 1st | 1840 | 1849 flood | Main Channel | Built by private subscriptions |
| Wells Street | 1st | 1841 | 1847 | Main Channel | Built by private subscriptions |
| Wells Street | 2nd | 1847 | 1849 flood | Main Channel | Boiler iron floats |
| Randolph Street | 2nd | 1847 | 1849 flood | South Branch | Semifloating draw |
| Madison Street | 1st | 1847 | 1849 flood | South Branch | Boiler iron floats and semifloating draw |
| Chicago Avenue | 1st | 1849 | 1867 | North Branch | Wood bent with 20' draw; 1850 petition to enlarge the draw was tabled |

## Pontoon-Turntable Swing Bridges

| Location | No. | Built | Removed | Waterway | Notes |
|---|---|---|---|---|---|
| Kinzie Street | 3rd | 1849 | 1859 | North Branch | Light turntable on piles with float draw |
| Clark Street | 2nd | 1849 | 1853 | Main Channel | Turntable with cast-iron balls |
| Wells Street | 3rd | 1849 | 1856 | Main Channel | |
| Randolph Street | 3rd | 1849 | 1856 | South Branch | Turntable design level with street |
| Madison Street | 2nd | 1849 | 1856 | South Branch | Turntable design level with street |
| Van Buren Street | 1st | 1850 | 1858 | South Branch | Built by D. Harper, turntable design level with street (Madison Street plan) |
| Polk Street | 1st | 1855 | 1856 | South Branch | Old Clark Street Bridge, turntable with cast-iron balls |
| 12th Street/Roosevelt Road | 1st | 1855 | 1868 | South Branch | One-third financed by the City, two-thirds by subscriptions |
| North Avenue | 1st | 1858 | 1865 | North Branch | Old Randolph Street Bridge, turntable design level with street |
| Polk Street | 2nd | 1856 | 1869 | South Branch | |

## Pontoon-Turntable Swing Bridges (continued)

| Location | No. | Built | Removed | Waterway | Notes |
|----------|-----|-------|---------|----------|-------|
| 18th Street | 1st | 1857 | 1867 | South Branch | Turntable design |
| Erie Street | 1st | 1860 | 1871 | North Branch | |
| 35th Street | 1st | 1871 | 1873 | South Branch | Old Erie Street Bridge |
| Weed Street | 2nd | 1905 | 1910 | North Branch | Formerly temporary bridge at North Western Avenue, moved to Wilson Avenue on July 26, 1910 |
| Torrence Avenue | 1st | 1905 | 1938 | Calumet River | Financed by the City |
| Wilson Avenue | 1st | 1910 | 1913 | North Branch | Old Weed Street Bridge |
| Ogden Avenue (river) | 1st | 1908 | 1933 | North Branch | Financed by the City |
| 130th Street | 1st | 1930 | 1944 | Calumet River | |
| S. Ashland Avenue | Temp. | 1907 | 1908 | South Fork of S.B. | Temporary bridge, formerly 22nd Street Bridge |

## Pivot and Swing Bridges

| Location | No. | Built | Removed | Waterway | Notes |
|----------|-----|-------|---------|----------|-------|
| Lake Street | 1st | 1852 | 1859 | South Branch | Built by D. Harper, first pivot bridge in Chicago |
| Clark Street | 3rd | 1854 | 1858 | Main Channel | Built by D. Harper, pivot, wood |
| Wells Street | 4th | 1856 | 1862 | Main Channel | Built by D. Harper, pivot, wood |
| Randolph Street | 4th | 1856 | 1864 | South Branch | Built by Stone, Boomer & Bouton, used ICRR type turntable |
| Rush Street | 1st | 1856 | 1863 | Main Channel | Financed one-third by City, G&CU, ICRR, first all-iron swing bridge in Chicago |
| Madison Street | 3rd | 1857 | 1875 | South Branch | First 100 percent City-financed bridge, swing, iron |
| Van Buren Street | 2nd | 1858 | 1867 | South Branch | Pivot, wood |
| Clark Street | 4th | 1858 | 1866 | Main Channel | Pivot, wood |
| Lake Street | 2nd | 1859 | 1867 | South Branch | Wood |
| Kinzie Street | 4th | 1859 | 1870 | North Branch | Wood |
| S. Halsted Street | 1st | 1861 | 1871 | South Branch | Wood |

## Pivot and Swing Bridges (continued)

| Location | No. | Built | Removed | Waterway | Notes |
|---|---|---|---|---|---|
| Courtland Avenue | 3rd | 1862 | 1873 | North Branch | Pivot, wood truss |
| Wells Street | 5th | 1862 | 1871 fire | Main Channel | Pivot, wood |
| Randolph Street | 5th | 1864 | 1874 | South Branch | Wood |
| Rush Street | 2nd | 1864 | 1871 fire | Main Channel | Swing, iron |
| Fuller Street | 1st | 1865 | 1876 | South Branch | Wood |
| State Street | 1st | 1865 | 1871 fire | Main Channel | Wood |
| North Avenue | 2nd | 1866 | 1877 | North Branch | Pivot, wood truss |
| N. Halsted Street (river) | 1st | 1866 | 1877 | North Branch | Wood |
| Clark Street | 5th | 1866 | 1871 fire | Main Channel | Financed by the City, designed by J. K. Thompson, wood & iron |
| Van Buren Street | 3rd | 1867 | 1871 fire | South Branch | Financed by the City, designed by J. K. Thompson, wood & iron |
| Lake Street | 3rd | 1867 | 1885 | South Branch | Financed by the City, designed by J. K. Thompson, wood & iron |
| 18th Street | 2nd | 1867 | 1888 | South Branch | Financed by the City, designed by J. K. Thompson, wood & iron |
| 12th Street/Roosevelt Road | 2nd | 1868 | 1887 | South Branch | Financed by the City, designed by J. K. Thompson, wood & iron |
| Chicago Avenue | 2nd | 1867 | 1871 fire | North Branch | Financed by the City, designed by J. K. Thompson, wood & iron |
| Throop Street | 1st | 1868 | 1903 | South Branch | Pivot, wood & iron |
| Grand Avenue | 1st | 1869 | 1910 | North Branch | Financed by the City, wood & iron |
| W. Division Street (river) | 1st | 1869 | 1902 | North Branch | Wood & iron |
| Adams Street | 1st | 1869 | 1871 fire | South Branch | Wood & iron |
| Polk Street | 3rd | 1869 | 1871 fire | South Branch | Wood & iron |
| S. Western Avenue (near 26th) | 1st | 1869 | 1899 | West Fork of S.B. | Wood & iron |
| Kinzie Street | 5th | 1870 | 1907 | North Branch | Financed by the City, wood & iron |
| Division Street (canal) | 1st | 1870 | 1900 | North Branch | Wood & iron |

## Pivot and Swing Bridges (continued)

| Location | No. | Built | Removed | Waterway | Notes |
|---|---|---|---|---|---|
| Archer Avenue | 3rd | 1870 | 1905 | South Branch | Wood & iron |
| Erie Street | 2nd | 1871 | 1908 | North Branch | Wood & iron |
| 22nd Street/Cermak Road | 1st | 1871 | 1905 | South Branch | Wood & iron |
| S. Ashland Avenue (W.F.) | 2nd | 1871 | 1883 | West Fork of S.B. | Wood |
| S. Ashland Avenue (S.F.) | 1st | 1871 | 1907 | South Fork of S.B. | Wood |
| Rush Street | 3rd | 1872 | 1883 | Main Channel | Iron |
| Clark Street | 6th | 1872 | 1889 | Main Channel | Financed by the City, wood & iron (moved to Webster St.) |
| S. Halsted Street | 2nd | 1872 | 1892 | South Branch | Financed by the City, iron |
| Wells Street | 6th | 1872 | 1888 | Main Channel | Iron |
| State Street | 2nd | 1872 | 1887 | Main Channel | Financed by the City, iron |
| Van Buren Street | 4th | 1872 | 1894 | South Branch | Wood & iron |
| Chicago Avenue | 3rd | 1872 | 1911 | North Branch | Wood & iron |
| Adams Street | 2nd | 1872 | 1889 | South Branch | Iron (moved to Taylor Street) |
| Polk Street | 4th | 1872 | 1907 | South Branch | Iron |
| 35th Street | 2nd | 1873 | 1891 | South Branch | Wood & iron |
| Courtland Avenue | 4th | 1873 | 1901 | North Branch | Wood & iron |
| Randolph Street | 6th | 1874 | 1903 | South Branch | Iron (1887, steam powered) |
| N. Halsted Street (canal) | 1st | 1874 | 1907 | North Branch | Iron |
| Madison Street | 4th | 1875 | 1891 | Main Channel | Iron (1887, steam powered) |
| Archer Avenue (Ogden Slip) | 1st | 1876 | 1896 | Ogden Slip off S.B. | Financed by the City, wood & iron |
| Fuller Street | 3rd | 1877 | 1909 | South Branch | Wood & iron |
| Harrison Street | 1st | 1877 | 1902 | South Branch | Iron |
| North Avenue | 3rd | 1877 | 1906 | North Branch | Wood & iron |
| N. Halsted Street (river) | 2nd | 1877 | 1896 | North Branch | Bow truss, wood |
| Fullerton Avenue | 2nd | 1877 | 1895 | North Branch | Two-thirds financed by the City, one-third by Lake View, wood & iron |
| S. Ashland Avenue (W.F.) | 3rd | 1883 | 1902 | West Fork of S.B. | Iron |

## Pivot and Swing Bridges (continued)

| Location | No. | Built | Removed | Waterway | Notes |
|---|---|---|---|---|---|
| Lake Street | 4th | 1886 | 1916 | South Branch | Iron |
| Rush Street | 4th | 1884 | 1920 | Main Channel | Iron, steam powered (1895 electrified) |
| 12th Street/Roosevelt Road | 3rd | 1887 | 1920 | South Branch | Steel, steam powered |
| State Street | 3rd | 1887 | 1901 | Main Channel | Steel, steam powered (1897 electrified) |
| 18th Street | 3rd | 1888 | 1903 | South Branch | Iron & steel (1894, received Lake Street steam engine) |
| Wells Street | 7th | 1888 | 1921 | Main Channel | Finance by North Chicago Street Ry., steel |
| Dearborn Street | 2nd | 1888 | 1907 | Main Channel | Old Wells Street Bridge, financed by the City, iron, steam powered |
| Jackson Street | 1st | 1888 | 1915 | South Branch | Steel, steam powered |
| Morgan Street | 1st | 1888 | 1913 | South Fork of S.B. | Iron |
| Loomis Street | 1st | 1888 | 1904 | South Branch | Wood & steel |
| Clark Street | 7th | 1889 | 1929 | South Branch | Steel |
| Adams Street | 3rd | 1889 | 1925 | South Branch | Steel |
| Webster Avenue | 1st | 1889 | 1914 | North Branch | Old Clark Street Bridge, wood & iron |
| Taylor Street | 1st | 1890 | 1899 | South Branch | Old Adams Street Bridge, iron |
| Canal Street | 1st | 1890 | 1892 | South Branch | Financed by the City, wood & iron (moved to Belmont Avenue) |
| N. Western Avenue | 1st | 1891 | 1903 | North Branch | Wood & iron on an iron & steel turntable |
| Madison Street | 5th | 1891 | 1922 | South Branch | Steel |
| Washington Street | 1st | 1891 | 1907 | South Branch | Iron |
| 35th Street | 3rd | 1891 | 1912 | South Branch | Howe truss |
| Belmont Avenue | 2nd | 1893 | 1911 | North Branch | Old Canal Street Bridge, Howe truss, wood & iron |
| Fullerton Avenue | 3rd | 1895 | 1960 | North Branch | Financed by the City & North Chicago Ry., steel |
| Diversey Avenue | 1st | 1896 | 1967 | North Branch | Financed by the City, steel, steam powered |
| Blackhawk Street | 1st | 1902 | 1910 | North Branch | Wood & iron |
| S. Ashland Avenue | 2nd | 1908 | 1920 | South Fork of S.B. | Steel |

## Harmon Folding Lift (Jackknife) Bridges

| Location | No. | Built | Removed | Waterway | Notes |
|---|---|---|---|---|---|
| Weed Street | 1st | 1891 | 1899 | North Branch | Financed by the City, steel |
| Canal Street | 2nd | 1893 | 1903 | South Branch | Financed by the City, steel |

## Vertical-Lift Bridges

| Location | No. | Built | Removed | Waterway | Notes |
|---|---|---|---|---|---|
| S. Halsted Street | 3rd | 1894 | 1931 | South Branch | Financed by the City |
| Torrence Avenue | 2nd | 1938 | Present | Calumet | Financed by the City |

## Scherzer Rolling-Lift Bridges

| Location | No. | Built | Removed | Waterway | Notes |
|---|---|---|---|---|---|
| Van Buren Street | 5th | 1895 | 1956 | South Branch | Financed by the City and the Metropolitan Railroad |
| Metropolitan Elevated Train Bridge | 1st | 1895 | 1958 | South Branch | Financed by the Metropolitan Railroad |
| N. Halsted Street (river) | 3rd | 1897 | 1955 | North Branch | Three-quarters financed by the City, one-quarter by the North Chicago Ry. |
| Taylor Street | 2nd | 1901 | 1929 | South Branch | Financed by the Sanitary District |
| Canal Street | 3rd | 1903 | 1941 | South Branch | Financed by the Sanitary District |
| Randolph Street | 7th | 1903 | 1981 | South Branch | Financed by the Sanitary District |
| S. Throop Street | 2nd | 1903 | 1978 | South Branch | Financed by the Sanitary District |
| State Street | 4th | 1903 | 1939 | Main Channel | Financed by the Sanitary District |
| Loomis Street | 2nd | 1904 | 1977 | South Branch | Financed by the Sanitary District |
| Harrison Street | 2nd | 1905 | 1959 | South Branch | Financed by the Sanitary District |
| 18th Street | 4th | 1905 | 1966 | South Branch | Financed by the Sanitary District |
| Cermak Road | 2nd | 1905 | Present | South Branch | Financed by the Sanitary District |
| Dearborn Street | 3rd | 1907 | 1959 | Main Channel | Financed by the Sanitary District |
| Pennsylvania RR "8-track" | 1st | 1910 | Present | Sanitary Canal | Financed by the Sanitary District; fixed bridges built in 1901 were converted to moveable in 1909–10 |

## Page Bascule Bridge

| Location | No. | Built | Removed | Waterway | Notes |
|---|---|---|---|---|---|
| S. Ashland Avenue | 4th | 1902 | 1936 | West Fork of S.B. | Financed by the Sanitary District |

## Strauss Trunnion Bascule Bridges

| Location | No. | Built | Removed | Waterway | Notes |
|---|---|---|---|---|---|
| Polk Street | 5th | 1910 | 1972 | South Branch | Financed by the City, pony truss, undermout external rack used in Milwaukee, Wis. |
| Jackson Street | 2nd | 1916 | Present | South Branch | Financed by the Sanitary District, below-deck truss |
| Lake Shore Drive | 1st | 1937 | Present | Main Channel | Built by the Chicago Park District, double deck |
| Lake Shore Drive, Ogden Canal | 1st | 1937 | 1982 | Michigan Canal | Built by the Chicago Park District, double deck |

## Chicago-Type Bascule Bridges—First Generation (1902–10)

| Location | No. | Built | Removed | Waterway | Notes |
|---|---|---|---|---|---|
| Courtland Avenue | 5th | 1902 | Present | North Branch | Financed by the City, thru-truss |
| 95th Street | 3rd | 1903 | 1958 | Calumet | Financed by the City, thru-truss |
| E. Division Street (canal) | 2nd | 1903 | Present | North Branch | Financed by the City, thru-truss |
| W. Division Street (river) | 2nd | 1904 | Present | North Branch | Financed by the City, thru-truss |
| N. Western Avenue | 2nd | 1904 | 1973 | North Branch | Financed by the City, thru-truss |
| Archer Avenue (single leaf) | 4th | 1906 | 1963 | South Fork of S.B. | Financed by the City, thru-truss |
| North Avenue | 4th | 1907 | 2006 | North Branch | Financed by the City, thru-truss |
| N. Halsted Street (canal) | 2nd | 1908 | 2011 | North Branch | Financed by the City, thru-truss |
| Kinzie Street (single leaf) | 6th | 1909 | Present | North Branch | Financed by the City, thru-truss |
| Erie Street | 3rd | 1910 | 1971 | North Branch | Financed by the City, pony truss |

## Chicago-Type Bascule Bridges—Second Generation: Improvement Period (1911–22)

| Location | No. | Built | Removed | Waterway | Notes |
|---|---|---|---|---|---|
| Grand Avenue | 2nd | 1913 | Present | North Branch | Financed by the City, pony truss |
| Washington Street | 2nd | 1913 | Present | South Branch | Financed by the City, pony truss |
| Chicago Avenue | 4th | 1914 | Present | North Branch | Financed by the City, pony truss |
| 35th Street (single leaf) | 4th | 1914 | 1969 | South Branch | Financed by the City, thru-truss |
| 92nd Street | 2nd | 1914 | Present | Calumet | Financed by the City, pony truss |
| Lake Street | 5th | 1916 | Present | South Branch | Double-deck thru-truss |
| Webster Avenue | 2nd | 1916 | Present | North Branch | Pony truss |
| Belmont Avenue | 3rd | 1917 | 1976 | North Branch | Pony truss |

## Chicago-Type Bascule Bridges—Second Generation: Improvement Period (1911–22) (continued)

| Location | No. | Built | Removed | Waterway | Notes |
|---|---|---|---|---|---|
| Monroe Street | 1st | 1919 | Present | South Branch | Financed by the Union Station Company, pony truss |
| Michigan Avenue | 1st | 1920 | Present | Main Channel | Double-deck thru-truss |
| Franklin-Orleans Avenue | 1st | 1920 | Present | Main Channel | Pony truss |
| Madison Street | 6th | 1922 | Present | South Branch | Rail-height pony truss |
| Wells Street | 8th | 1922 | Present | Main Channel | Double-deck thru-truss |

## Chicago-Type Bascule Bridges—Second Generation: Refinement Period (1923–41)

| Location | No. | Built | Removed | Waterway | Notes |
|---|---|---|---|---|---|
| California Avenue | | 1926 | Present | Sanitary & Ship Canal | Financed by the Sanitary District, pony truss |
| Adams Street | 4th | 1927 | Present | South Branch | Financed by the City, below-deck truss |
| 100th Street | 1st | 1927 | Present | Calumet | Pony truss |
| LaSalle Street | 1st | 1928 | Present | Main Channel | Financed by the City, pony truss |
| N. Damen Avenue (single leaf) | 1st | 1929 | 1998 | North Branch | Financed by the City, below-deck truss |
| Clark Street | 8th | 1929 | Present | Main Channel | Financed by the City, pony truss |
| 106th Street | 2nd | 1929 | Present | Calumet | Pony truss |
| Roosevelt Road | 4th | 1930 | Present | South Branch | Financed by the City, below-deck truss |
| S. Damen Street | 1st | 1930 | 2001 | Sanitary & Ship Canal | Financed by the City |
| Wabash Street | 1st | 1930 | Present | Main Channel | Financed by the City, rail-height pony truss |
| Ogden Avenue (canal) | 1st | 1932 | 1994 | North Branch | Financed by the City |
| Ogden Avenue (river) | 2nd | 1932 | 1994 | North Branch | Financed by the City |
| S. Halsted Street | 4th | 1934 | Present | South Branch | Triple-pony truss |
| N. Ashland Avenue | 1st | 1936 | Present | North Branch | Financed by the Board of Local Improvements, pony truss |
| S. Ashland Avenue | 5th | 1938 | Present | West Fork of S.B. | Triple-pony truss, replaced Page bascule bridge |

## Chicago-Type Bascule Bridges—Post–World War II (1948–67)

| Location | No. | Built | Removed | Waterway | Notes |
|---|---|---|---|---|---|
| Canal Street | 4th | 1948 | Present | South Branch | Pony truss, two bridge houses, replaced Scherzer rolling lift |
| State Street | 5th | 1949 | Present | Main Channel | Rail-height pony truss, two bridge houses, replaced Scherzer rolling lift |
| N. Halsted Street (river) | 4th | 1955 | Present | North Branch | Below-deck truss, one bridge house, replaced Scherzer rolling lift |
| Van Buren Street | 6th | 1956 | Present | South Branch | Rail-height pony truss, one bridge house, replaced Scherzer rolling lift |
| Congress Parkway (twin) | 1st | 1956 | Present | South Branch | Financed by City/state/federal funds, two bridge houses |
| 95th Street | 2nd | 1958 | Present | Calumet | Pony truss, one bridge house, replaced Scherzer rolling lift |
| Harrison Street | 3rd | 1960 | Present | South Branch | Rail-height pony truss, one bridge house, replaced Scherzer rolling lift |
| Ohio Street (twin) | 1st | 1961 | Present | North Branch | Financed by City/state/federal funds, below-deck truss, two bridge houses |
| Dearborn Street | 4th | 1963 | Present | Main Channel | Rail-height pony truss, one bridge house, replaced Scherzer rolling lift |
| 18th Street (single leaf) | 5th | 1967 | Present | South Branch | Pony truss, one bridge house, replaced Scherzer rolling lift |

## Chicago-Type Bascule Bridges—Modern

| Location | No. | Built | Removed | Waterway | Notes |
|---|---|---|---|---|---|
| Loomis Street | 3rd | 1978 | Present | South Branch | Financed by the City, rail-height pony truss, steel box-girder design |
| Columbus Drive | 1st | 1982 | Present | Main Channel | Financed by the City, deck truss, steel box-girder design |
| Randolph Street | 7th | 1984 | Present | South Branch | Financed by state and federal funds, rail-height pony truss, steel box-girder design |

# NOTES

## INTRODUCTION

1. Gurdon S. Hubbard, *Incidents and Events in the Life of Gurdon Saltonstall Hubbard,* 42–43.

2. "Drainage Board Adopts Committee Report for $2,500,000 Bond Issue," *Chicago Daily Tribune,* September 13, 1900, 1.

3. Chicago Bureau of Engineering, Division of Bridges and Viaducts, *Chicago River Survey,* 5.

4. Ulrich Danckers and Jane Meredith, *Early Chicago: A Compendium of the Early History of Chicago to the Year 1835 When the Indians Left,* 84.

5. Alfred Andreas, *History of Chicago: From the Earliest Period to the Present Time,* 1:197. The table is reprinted without correction by the author.

6. Ibid., 106.

7. Untitled article, *Chicago Daily American,* April 17, 1839, 2 (microfilm, Research Center, Chicago History Museum, Chicago, 1835–42).

8. Edwin O. Gale, *Reminiscences of Early Chicago and Vicinity,* 293–94.

9. Charles Cleaver, Esq., "Reminiscences of Early Chicago (1833)," 52.

10. James K. Thompson, improved turntable, U.S. Patent 78,553, dated June 2, 1868; truss bridge, U.S. Patent 81,960, dated September 8, 1868.

11. "Keep Off the Bridge," *Chicago Daily Tribune,* August 21, 1856, 3.

12. City of Chicago, *Thirteenth Annual Report of the Board of Public Works,* 70.

13. *Escanaba & Lake Michigan Transportation Company v. City of Chicago,* 107 U.S. 678 (1883).

14. Citizens Association of Chicago, Committee on Street Railways and Bridges, *Report of the Committees on Bridges and Street Railways of the Citizens' Association of Chicago,* 3.

15. J. A. L. Waddell, lift bridge, U.S. Patent 506,571, filed November 10, 1892, and issued October 10, 1893.

16. Joseph B. Strauss, bridge (bascule), U.S. Patent 668,232, filed May 31, 1900, and issued February 19, 1901; bridge (bascule), U.S. Patent 738,954, filed December 19, 1902, and issued September 15, 1903.

17. Alexander F. L. von Babo, trunnion bascule bridge, U.S. Patent 1,001,800, filed June 26, 1908, and issued August 29, 1911.

18. U.S. Court of Appeals, *The Scherzer Rolling Lift Bridge Company v. City of Chicago and Great Lakes Dock Company,* Seventh Circuit, Records and Briefs, Case No. 3606, in Record Group 276, National Archives, Chicago, October 1924.

19. U.S. Department of the Interior, *Historic American Engineering Record,* "Addendum to Chicago River Bascule Bridge, Monroe Street," 5.

20. City of Chicago, Department of Public Works, *Fifty-Fourth Annual Report of the Department of Public Works for the Year Ending December 31, 1929,* 216–17.

## THE BRIDGES OF THE MAIN CHANNEL

### Rush Street Bridges

1. "The Rush Street Bridge Destroyed," *Chicago Tribune,* November 4, 1863, 4.

2. "Rush Street Bridge," *Chicago Daily Tribune,* November 24, 1883, 8.

### Wabash Avenue Bridge

1. "New Wabash Ave. Bridge Is Awarded Prize for Beauty," *Chicago Daily Tribune,* June 16, 1931, 17.

### State Street Bridges

1. "State Street Bridge," *Chicago Tribune,* August 21, 1862, 4.

2. "State Street Bridge," *Chicago Tribune,* December 8, 1863, 4.

3. Keystone Bridge Company, *Descriptive Catalogue of Wrought-Iron Bridges,* 33.

4. City of Chicago, Department of Public Works, *Tenth Annual Report of the Department of Public Works, Fiscal Year Ending December 31, 1885,* 70.

5. City of Chicago, Department of Public Works, "Report on Necessary Bridge Work," 21–22.

### Dearborn Street Bridges

1. Untitled article, *Chicago Daily American,* February 25, 1840, 3 (microfilm, Research Center, Chicago History Museum, Chicago, 1835–42).

2. "Bridge Almost a Wreck," *Chicago Daily,* May 24, 1902, 4.

### Clark Street Bridges

1. Andreas, *History of Chicago,* 198–99.

2. Untitled article, *Chicago Daily Tribune,* September 9, 1853, 3.

3. Untitled article, *Chicago Daily Press,* May 26, 1855, 3.

4. "Rebuild the Clark Street Bridge," *Chicago Press and Tribune,* July 29, 1858, 1.

5. Untitled article, *Chicago Press and Tribune,* March 4, 1859, 1.

6. "Clark Street Bridge," *Chicago Daily Tribune,* April 6, 1866, 2.

7. City of Chicago, Department of Public Works, *Thirteenth Annual Report of the Department of Public Works, for the Fiscal Year Ending December 31, 1888,* 208.

### LaSalle Street Bridge and Tunnel

1. "Officials Seek to Prevent Building of Ugly Bridges," *Chicago Daily Tribune,* September 13, 1930, 4.

### Wells Street Bridges

1. U.S. Department of the Interior, *Historic American Engineering Record,* "Addendum to Chicago River Bascule Bridge, Wells Street Bridge," 7–8.

2. Untitled article, *Chicago Tribune,* June 25, 1863, 4.

3. "Terrible Accident," *Chicago Tribune,* October 4, 1869, 4.

### Franklin-Orleans Street Bridge

1. U.S. Department of the Interior, *Historic American Engineering Record,* "Addendum to Chicago River Bascule Bridge, Franklin-Orleans Street," 5.

## THE BRIDGES OF THE SOUTH BRANCH

### Lake Street Bridges

1. City of Chicago, "Report of Committee on Harbor and Bridges on Petition of S. Lind, et al., to Operate the Lake St. Bridge at Night," 1.

2. City of Chicago, "Order to the Superintendent to Purchase and Drive Piles to Protect the Lake St. Bridge," 1.

3. "The City; Another Bridge Accident," *Chicago Tribune,* July 7, 1863, 4.

4. "Suicides: A Leap into the River," *Chicago Tribune,* July 27, 1871, 1.

5. "The New Lake Street Bridge," *Chicago Daily Tribune,* June 13, 1886, 18.

6. City of Chicago, Department of Public Works, *Eighteenth Annual Report of the Department of Public Works, Fiscal Year Ending December 31, 1893,* 43.

7. "Federal Government Wants Two River Bridges Removed," *Chicago Daily Tribune,* April 16, 1909, 1.

### South Branch Bridge

1. City of Chicago, "Petition for a Bridge at Lake St.," 1.

2. City of Chicago, "Proposal of J. F. Brown to Build a Bridge across the South Branch," 1.

3. Untitled article, *Chicago Democrat,* April 13, 1836, 2 (microfilm, Research Center, Chicago History Museum, Chicago, 1833–61).

4. City of Chicago, "Report of Committee on Streets and Bridges on the South Branch Bridge," 1.

### Randolph Street Bridges

1. City of Chicago, "Report of Special Committee on the Randolph St. Bridge Announcing Its Completion," 1.

2. "The Bridges," *Chicago Tribune,* May 11, 1864, 4.

3. "Photo Standalone 4—Untitled," *Chicago Tribune,* October 25, 1984, B3.

### Madison Street Bridges

1. Thomas W. H. Moseley, truss bridge, U.S. Patent 16,572, February 3, 1857.

### Adams Street Bridges

1. "Wanted, a Bridge," *Chicago Tribune,* December 23, 1866, 4.

### Jackson Boulevard Bridges

1. "Hand Operated Bridge Barriers Sought by Jury," *Chicago Daily Tribune,* March 27, 1937, 14.

### Van Buren Street Bridges and Tunnel

1. "The Flood!," *Chicago Daily Tribune,* February 9, 1857, 1.

2. William Scherzer, lift bridge, U.S. Patent 511,713, filed May 29, 1893, and issued December 26, 1893.

### Harrison Street Bridges

1. "Another Bridge Project," *Chicago Tribune,* March 9, 1872, 5.

### Polk Street Bridges

1. "Our Bridges," *Chicago Tribune,* May 13, 1864, 4.

2. "2 Bridges Set for Demolition," *Chicago Tribune,* May 27, 1971, S4.

### Taylor Street Bridges

1. "Want a Clear Channel," *Chicago Daily,* July 22, 1898, 7.

### Roosevelt Road (Formerly 12th Street) Bridge

1. "The City's Interests," *Chicago Daily Tribune,* February 26, 1886, 6.

2. City of Chicago, Department of Public Works, *Eleventh Annual Report of the Department of Public Works, for the Year Ending December 31, 1886,* 148.

### 18th Street Bridges

1. "The City's Interests," *Chicago Daily Tribune,* February 26, 1886, 6.

2. City of Chicago, Department of Public Works, *Third Annual Report of the Department of Public Works for the Year Ending December 31, 1878,* 103.

### Canal Street Bridges

1. "Against the Canal Street Bridge," *Chicago Daily Tribune,* January 13, 1891, 3.

2. "Vigorous Protest Made," *Chicago Daily,* July 10, 1891, 12.

3. William Harman, bridge, U.S. Patent 383,880, filed March 19, 1888, and issued June 5, 1888.

4. "Laugh at the City," *Chicago Tribune,* June 3, 1894, 11.

5. City of Chicago, Department of Public Works, *Twenty-First Annual Report of the Department of Public Works for the Year Ending December 31, 1896,* 110.

### South Halsted Street Bridges

1. City of Chicago, Board of Public Works, *Eleventh Annual Report of the Board of Public Works, for the Municipal Year Ending March 31, 1872,* 43.

2. "Halsted Street Bridge Disabled," *Chicago Daily Tribune*, November 10, 1882, 3.

3. "Seven Wonders of Chicago," *Chicago Daily Tribune,* March 22, 1908, F4.

4. "Trapped in Midair," *Chicago Daily Tribune,* July 17, 1894, 8.

5. "Life Bridge Breaks Down," *Chicago Daily Tribune,* October 3, 1899, 10.

### Loomis Street Bridges

1. Michael K. Hopson, *The Comprehensive Guide to the Bridges of the Chicago River,* 154.

2. "Shot in the Chase," *Chicago Daily Tribune,* June 13, 1895, 8.

3. "City Submits Plan for New Loomis Bridge," *Chicago Daily Tribune,* May 24, 1970, S5.

### South Ashland Avenue Bridges

1. John W. Page, bascule bridge, U.S. Patent 731,322, filed November 13, 1902, and issued June 16, 1903.

2. City of Chicago, Department of Public Works, *Fiftieth Annual Report of the Department of Public Works for the Fiscal Year Ending December 31, 1925,* 166–17.

## THE BRIDGES OF THE NORTH BRANCH

### Kinzie Street Bridges

1. City of Chicago, "Report of Committee on Harbor & Bridges on Petition of John Rodgers, et al., for a New Bridge at Kinzie St.," 1–2.

2. Ibid., 1–2.

3. "Chicago River on Fire," *Chicago Daily,* April 18, 1899, 1.

4. Ibid.

5. City of Chicago, Department of Public Works, *Mayor's Annual Message and Thirtieth Annual Report of the Department of Public Works for the Year Ending December 31, 1905,* 264.

### Erie Street Bridges

1. "Bridge Collapses; A Wreck," *Chicago Daily Tribune,* May 19, 1908, 2.

2. "2 Bridges Set for Demolition," *Chicago Tribune,* May 27, 1971, W10.

### Chicago Avenue Bridges

1. City of Chicago, "Report Committee on Harbor & Bridges on Petition of A. Crosby and Co., et al. to Widen the Draw of the North Branch Bridge," 1.

2. "Orders Removal of Three Bridges," *Chicago Tribune*, January 29, 1911, 1.

### North Halsted Street Bridges (River)

1. City of Chicago, Department of Public Works, *First Annual Report of the Department of Public Works, for the Fiscal Year Ending December 31, 1876,* 164.

2, City of Chicago, Department of Public Works, *Mayor's Annual Message and Thirty-Fifth Annual Report of the Department of Public Works to the City Council of the City of Chicago for the Year Ending December 31, 1910,* 166.

3. "Auto Climbs Open Jackknife Bridge: Hurls Man Where?," *Chicago Daily Tribune,* September 12, 1913, 1.

4. Ibid. Several witnesses saw someone or something thrown from the car just before it stopped, but no one heard or saw any splash in the river. Police dragged the river for a body, but nothing more was ever discovered.

### North Halsted Street Bridges (Canal)

1. City of Chicago, Department of Public Works, *Mayor's Annual Message and Thirtieth Annual Report,* 263.

### East Division Street Bridges (Canal)

1. "Car Hangs over River's Edge," *Chicago Daily Tribune,* March 10, 1899, 1.

### Weed Street Bridges

1. "Works Like a Knife," *Chicago Daily Tribune,* June 26, 1891, 3. A counterweighted cam connected to a cable aided in the raising of the bridge, which required some force to begin the opening process but became perfectly balanced once in the open position.

2. "Bridges Leading to Goose Island, Whose People Fear They Soon Must Swim the River," *Chicago Daily Tribune,* January 26, 1901, 9.

### North Avenue Bridges

1. City of Chicago, "Report Committee on Harbor & Bridges on Petition of Henry Smith, et al., to Erect a Floating Bridge at North Ave.," 1.

2. City of Chicago, Board of Public Works, *Fourth Annual Report of the Board of Public Works to the Common Council of the City of Chicago, for the Municipal Fiscal Year Ending April 1, 1865,* 26.

### Courtland Street Bridges

1. "The Capture of the Clybourne Avenue Bridge," *Chicago Daily Tribune,* April 29, 1862, 4.

2. Ibid.

3. Ibid.

4. "Improvements on the North Branch," *Chicago Daily Tribune,* June 24, 1862, 4.

5. Ibid.

### North Ashland Avenue Bridge

1. "New $1,700,000 Ashland Avenue Bridge Is Open," *Chicago Daily Tribune,* August 21, 1936, 11.

### North Damen Avenue Bridges

1. "Thousands See Damen Avenue Bridge Opened," *Chicago Daily Tribune,* January 25, 1929, 14.

### Diversey Avenue Bridges

1. "Strengthened Diversey Bridge to Be Reopened," *Chicago Daily Tribune,* January 27, 1963, 6.

### Belmont Avenue Bridges

1. Thomas E. Leonard (former owner, Henry C. Grebe & Co.; owner, Grebe Consulting), in discussion with the author, March 1, 2007.

# GLOSSARY

**A. F. Shuman pavement**  A road surface patented by A. F. Shuman and produced by his company at 88 Washington Street in Chicago, Illinois. Shuman offered bituminous wood-plank sectional paving products in the first decade of the twentieth century. The company later offered wood and asphalt paving blocks, sometimes referred to as Shuman "W and A" pavement, and a corrugated surface pavement designed for inclines called "Slip-Not."

**abutment**  A retaining wall supporting the ends of a bridge or viaduct.

**Alundum**  A trademarked brand name for a hard infused alumina most frequently used as an abrasive or a refractory material to line furnaces.

**approach**  The section of the roadway connecting the street and the bridge deck.

**arch truss**  Typically identified by a curved structural member spanning an opening and providing significant support to the structure.

**bark**  Refers to any vessel with three or more masts that carried primarily square sails, excepting the fore and aft sails, which were usually triangular. The advantage of these ships was that they could operate with smaller crews, were cheaper to build, and combined the best features of a fully rigged ship and a schooner.

**baroque**  The building style of the baroque era that began in Italy in the late sixteenth century and ended in the eighteenth century. This style combined Roman and Renaissance architecture to create a new, freer use of classical orders and ornamentation. The resulting forms are characterized by new explorations of form, light, and shadow and dramatic intensity created to celebrate movement and triumph.

**bascule**  Originally a French term meaning "seesaw," this word comes from the noun *baculer,* meaning to strike or land on one's buttocks. It is now used to refer to the general moveable-bridge design that opens vertically by rotating or rocking back. The rising deck is usually balanced by lowering a counterweight, a design term first associated with the medieval drawbridge.

**bearing**  A housing or device that provides support and allows rotation. On bascule bridges, the trunnion is held by a bearing (or housing) to support and enable the smooth rotation of the bridge leaf. On a fixed bridge, the bearing sits on the top of a pier or abutment, and the ends of the bridge rest on it.

**Beaux Arts**  Refers to the neoclassical architectural style taught by the Paris School of Fine Arts (or École des Beaux-Arts) that began in the late seventeenth century and carried on into the early twentieth century. It heavily influenced architecture in the United States from the 1880s to the 1920s.

**Bedford limestone**  A regional term for Salem limestone used to describe a high-quality rock of south-central Indiana. It is mostly made up of calcium carbonate from decomposed marine fossils deposited over millions of years at the bottom of a shallow inland sea.

**bedrock**  The solid rock layer beneath sand or silt.

**bent**  A rigid frame commonly made of logs, wood beams, reinforced concrete, or steel that supports a vertical load. It is often used to support bridge superstructures, crossbeams, or girders. An *end bent* is the supporting frame forming part of an abutment. Each vertical member of a bent may be called a column, pier, or pile. The horizontal member resting on top of the columns is a bent cap.

**Bessemer process**   Patented in 1855 by Henry Bessemer, the first inexpensive industrial process for mass production of steel from molten pig iron.

**box girder**   Typically rectangular or trapezoidal in shape; made of prestressed concrete, structural steel, or a composite of steel and reinforced concrete. Box-girder bridges are commonly used for highway flyovers or for modern elevated structures of light-rail transport. Although normally used on beam bridges, box girders may also be used on cable-stayed bridges and bascule bridges.

**bridge deck**   The roadway or bridge floor system used by street, rail, or pedestrian traffic to cross a bridge.

**bridge operator**   The men and women employed to open and close drawbridges. Additional responsibilities may also include keeping a log and basic bridge and bridge-house maintenance. Originally called bridge tenders until the 1990s, in contract negotiations with the City of Chicago the union initiated this preferred term of *bridge operator.*

**Bridgeport**   The neighborhood on Chicago's South Side that takes its name from the junction of the Illinois & Michigan Canal and the South Branch of the Chicago River. Bridgeport is bound on the north and west by the Chicago River, on the south by Pershing Road, and on the east by Canal Street.

**bridge tender**   *See* bridge operator.

**Bubbly Creek**   The nickname given to the South Fork of the South Branch of the Chicago River. To this day, bubbles caused by the decomposition of organic waste that was regularly dumped into the river from the Union Stock Yards and related industries continue to rise to the surface.

**cable-stayed bridge**   A variation on the suspension bridge, with the bridge deck carried by cable tension members extending from one or more towers at varying angles. The design allows great freedom of form. It does not use cables draped over towers with anchorages at each end like in a traditional suspension bridge. The current North Avenue Bridge is a good example.

**caisson**   A box or structure designed to prevent entry of water, often through the use of internal air pressure. Caissons are usually opened at the bottom to allow excavation or underwater work.

**cantilever**   Any rigid structural member extending horizontally from its vertical support. These are typically used as structural elements on bridges and buildings. Many Chicago bridges have cantilevered sidewalks that extend off the sides of the bridge superstructure.

**center pier**   The circular pier of a swing bridge, typically positioned at the center of a waterway or river channel, that supports the bridge turntable and superstructure. The center pier is typically extended by a bridge protection pier to deflect collisions and prevent damage from passing ships.

**Chicago Department of Transportation**   Department responsible for public-way infrastructure, including planning, design, construction, maintenance, and management of city bridges, viaducts, street lighting, and traffic lights.

**Chicago Sanitary and Ship Canal**   Originally called the Chicago Drainage Canal, it reversed the flow of the Chicago River in 1900 by connecting to the Illinois River. Opened to ship traffic in 1911, it is the only connection between the Great Lakes and Mississippi watersheds following the closing of the Illinois & Michigan Canal in 1933. It was designed to keep the waste and runoff of the Chicago River from polluting Lake Michigan, the city's source of drinking water.

**chord**   Either of the two principal members of a truss extending from end to end and connected by members (usually referred to as the top and bottom chords).

**chord covering**   The exterior covering of tin or iron used to cover the chords of wood bridges and protect them from the elements (primarily moisture) so as to extend the life of a bridge. It was designed to protect untreated or unpainted wood from dry rot. The tin and iron chord coverings regularly needed renewal because the sulfuric acid in the smoke from coal-burning tugs and steamships would combine with moisture in the atmosphere to create the sulfuric acid that would cause pitting and corrosion of the chord covers.

**City**   When capitalized, refers to the government of the City of Chicago; when lowercase, usually refers to Chicago in general.

**combination bridge**   A term used in the mid- to late 1800s that refers to a type of common wood and iron bridge. Usually Howe-truss bridges, they had wood top and bottom chords and interconnecting wood members, with iron tension rods running vertically through the trusses and bolted top and bottom.

**Common Council**   The precursor to the Chicago City Council.

**compression member**   A timber or other truss member that is sub-

jected to squeezing or pushing. Some members may act in both a compression and a tension capacity, particularly under live loads. *See also* tension.

**Copperhead Party**   A vocal group of northern state Democrats who opposed the American Civil War and argued for a peace settlement with the Confederate States.

**counterweight**   A weight used as a counterbalance.

**counterweight pit**   An open chamber, typically below street grade, incorporated into a bascule-bridge foundation that allows the short arm (or back end) of a bridge leaf and counterweight to rotate down into the pit to open the bridge.

**deck truss**   A bridge supported by one or more trusses beneath the roadway.

**double deck**   Two decks situated one above the other. This term is often used in reference to a bascule bridge, but refers to any bridge that conveys a combination of street, rail, or pedestrian traffic on two different levels.

**double leaf**   Two leaves, or moveable roadways. This often refers to a bascule bridge that has two leaves, usually meeting in the center of the bridge, that form a continuous bridge roadway.

**draw**   The opening presented by a moveable bridge. It may also refer to the floating pier, which is drawn to the side to open a pontoon swing bridge.

**dredge**   To catch, gather, or remove mud, sand, silt, and the like from the bottom of a body of water. This also refers to the various machines or vessels employed in dredging earth from a body of water.

**École des Beaux Arts**   School of Fine Arts in Paris, France.

**eyebars**   Iron or steel rods with a forged, machined, or attached end fitting, with a hole on one or both ends designed to accept a connecting pin. Around the turn of the twentieth century, pin connections were a popular method of connecting the truss members of a bridge.

**folding-lift bridge**   A bascule-bridge design patented by Captain William Harman, in which each leaf is hinged at its center, pivots at the shore, and is raised up at the middle and folded back upon itself, similar to a jackknife. Harman received U.S. Patent No. 383,880 for his design on June 5, 1888.

**freshnet**   A flood.

**friction rollers**   Balls or wheels that receive the pressure or weight of bodies in motion and relieve friction. These are usually combined with a center pin or pivot that carries the majority of the weight.

**frieze**   Any decorative band on an outside wall. Broader than a string-course, friezes often bear lettering, sculpture, or architectural relief.

**gallows frame**   A frame from which criminals were executed by hanging in the nineteenth century, usually consisting of two upright posts and a crossbeam on the top; also, a like frame for suspending anything. The term was used to refer to the rectangular frame used at each approach of Chicago's first drawbridge at Dearborn Street.

**gear train**   A connected set of rotating gears by which force is transmitted or motion or torque is changed.

**girder**   A horizontal structure member that supports vertical loads by resisting bending. Typically, it is a large beam, especially when made of multiple metal plates. The plates are usually riveted or welded together.

**Goose Island**   An artificial island on the North Branch that was named after the geese that were kept there by Irish immigrants during the late 1800s.

**gusset plate**   A plate that is used to connect beams and girders to a column or to connect members of a truss. A gusset plate is fastened to the other members through welding, riveting, bolting, or a combination of the three.

**gyroscopic controls**   Limiting controls that use a gyroscope to stop or slow the motion of a bridge or system at its farthest ranges of travel. On Chicago-type bascules, these controls were used to automatically slow and then stop the electric drive motors of the bridges as they approached the fully open or fully closed positions.

**hand-haul**   *See* warp.

**heel-trunnion**   A design in which the point of rotation, or trunnion, is located at the end, or heel, of an object. Patented by Joseph Strauss, heel-trunnion bascule bridges were used by many American railroad companies in the first half of the twentieth century.

**hipped roof**   A type of roof where all sides of the roof slope downward to the walls, usually gently, so there are no gables or other vertical sides to the roof. Each face of a hipped roof typically has the same pitch or slope; thus, hipped roofs are symmetrical about their centerlines. A square hipped roof is shaped like a pyramid. Hipped roofs have a consistent, level fascia, meaning that a gutter can be fitted all around.

**interlocking**   A method of tying multiple operations or electrical switches to a system that prevents successfully activating any switch or operation out of sequence.

**jackknife bridge**   The nickname used for the Harman folding-lift bascule bridge first built at Weed Street; later used to refer to subsequent bascule bridges such as the Scherzer rolling-lift or early Chicago-type bascule bridge. *See also* folding-lift bridge.

**jibboom**   A spar that forms an extension of the bowsprit of a ship.

**keep**   The innermost and strongest structure or central tower of a medieval castle.

**key**   A steel or iron bar, usually four to six feet in length, employed to manually open a swing bridge. One end of the key mates to a fitting near the center of the bridge deck. Walking the bar around in a circle turns a drive gear within the turntable to rotate the bridge.

**lateral bracing**   Members used to stabilize a structure by introducing diagonal connections.

**leaf**   The moving portion of a bascule bridge that includes the roadway, typically consisting of a long and short arm defined by the point of rotation, and often incorporating a counterweight. Bascule bridges are either single- or double-leaf bridges.

**"L" train**   Short for *elevated* train. Commonly used to refer to the elevated commuter trains operated by the Chicago Transit Authority and earlier commuter rail companies that operated in Chicago, Illinois.

**mansard roof**   A specific variation of a hipped roof, in which each side has two slopes and the lower slope is at a steeper angle than the upper slope. François Mansart (1598–1666), an accomplished architect of the French baroque period, first popularized this type of roof.

**member**   An individual angle, beam plate, or built piece intended to be an integral part of an assembled frame or structure.

**Metropolitan Water Reclamation District**   Formally the Metropolitan Water Reclamation District of Greater Chicago, usually abbreviated MWRD and originally called the Sanitary District of Chicago. It was chartered in 1889 by the State of Illinois to handle water sanitation and to protect Lake Michigan, the source of drinking water for the city of Chicago. This was accomplished through the reversal of the Chicago River in 1900 and through the construction and opening of the Chicago Sanitary and Ship Canal. The district is funded through area property taxes and manages wastewater treatment for the Chicago metropolitan area.

**Michigan Canal**   This fifteen-hundred-foot-long slip runs east and west parallel to the Chicago River and passes under Lake Shore Drive near Navy Pier. The Chicago Dredge and Dock Company dug this canal between 1902 and 1920 inside the locks separating Lake Michigan and the Chicago River to serve the Pugh Terminal (now known as the North Pier). Due to William Ogden's significant involvement with the company, this slip was often referred to as the Ogden Canal or Ogden Slip. Another slip by the same name also existed off of the South Branch of the Chicago River. *See also* Ogden Slip.

**Mud Lake**   Also known as Portage Lake. A wet, swampy lowland area, approximately six miles long, that extended along the Chicago Portage from a mile west of the West Fork of the South Branch to an eastern bend in the Des Plaines River.

**neoclassical**   An architectural style produced by the neoclassical movement that began in the mid–eighteenth century derived principally from classical Greek architecture. It was the style of architecture used for the "White City" in the World's Columbian Exposition of 1893, as overseen by architect Daniel Burnham.

**Nicholson pavement**   Patented by Samuel Nicholson of Boston in 1866, this road surface consisted of pine blocks three to four inches wide, six to fourteen inches long, and six inches deep, soaked in creosote. Abundant timber made this affordable. The surface was found to be slippery when wet and offered poor traction on inclines. The technique fell out of favor in Chicago after 1871, as the prevailing view was that Nicholson pavement had contributed to the rapid spread of the Great Chicago Fire.

**obstruction to navigation**   A phrase used in the Rivers and Harbors Act of 1890 to describe structures that impeded or prevented the navigation of national waterways. These structures were subject to federal review, and authority was given to the U.S. War Department to force removal. This duty is now overseen by the U.S. Coast Guard.

**Ogden Canal**   *See* Michigan Canal. *See also* Ogden Slip.

**Ogden Slip**   A slip that projected southeast from the South Branch, near Green Street, that was crossed by the Chicago Alton and St. Louis Railroad and by Archer Avenue via a shared swing bridge from 1876 to 1896. It was originally referred to as Healey's Slough. Ogden Slip was also used to refer to another slip that parallels the Chicago River at the lakefront. *See also* Michigan Canal.

**one-person operation**   Refers to the conversion of bascule bridges during the 1950s and '60s from a single bridge house, which was originally referred to as one-man operation. New technology allowed remote control of both sets of operating machinery on the drawbridges from one bridge house. For bridges on the Chicago River, this method of operation is typically bypassed for bridges with two bridge houses, as a bridge operator in each will control one leaf of the bridge with use of the leap-frog bridge crew system for the spring and fall runs.

**Page bascule**   A bascule-bridge design developed and patented in 1903 by John Page, a former Sanitary District engineer.

**parapet**   A low wall along the outside edge of a bridge deck, used to protect vehicles and pedestrians.

**pier**   A raised structure, typically supported by widely spread piles, pilings, or pontoons. Piers vary widely in size and structure. *See also* pile or piling.

**pile driver**   A machine that repeatedly drops a heavy weight on top of a pile; used to push a pile into the earth.

**pile or piling**   A long column, often wooden, driven deep into the ground to form part of a foundation or substructure. Often used as a key component of a pier or bridge protection on or along a waterway. *See also* pier.

**pin**   A cylindrical bar used to connect various members of a truss, such as those inserted through the holes of a meeting pair of eyebars.

**pinion**   A gear with a small number of teeth, especially one that engages with a rack or larger gear.

**pivot**   A pin, point, or short shaft on which something rests, turns, or rotates.

**pontoon**   A boat or some other floating structure used as one of the supports for a temporary bridge or pier.

**pony truss**   A truss extending above the bridge deck in which the top chord is not joined to the top chord of another truss by overhead braces.

**portage**   The carrying of boats, goods, and more over land from one waterway to another; also the name for the route by which this is done.

**Portage Lake**   *See* Mud Lake.

**propeller ships**   A term used to describe prop-driven ships.

**protection pier**   A series of pilings, often including piling clumps at the corners, tied together by caps, wales, or timbers. Typically used to protect an open swing bridge from collisions.

**pylon**   A monumental vertical structure marking the entrance to a bridge or the forming part of a gateway.

**rack**   A bar, with teeth on one of its sides, adapted to engage with the teeth of a pinion (rack and pinion). Used for converting circular into linear motion or vice versa.

**rail-height truss**   A type of pony truss in which the steel structure running above the bridge roadway is the approximate height of the railing.

**railing**   A fence-like construction built at the outermost edge of the roadway or the sidewalk portion of a bridge to protect pedestrians and vehicles.

**reinforced concrete**   Concrete with steel bars or mesh embedded for increased strength in tension.

**rim-bearing turntable**   A turntable in which the distribution of weight is supported by a ring around its outside circumference. This is typically accomplished through the use of a drum resting on a series of rollers that allow the circular rim of the turntable, which is then usually enclosed by a large circular rack or gear, to rotate. The best rim-bearing turntables distribute the weight of the bridge superstructure equally over eight symmetrical points on the bridge drum. Under this system, the rollers, and the tread on which they roll, should wear evenly. The ideal turntable design combines rim and center bearing; by carrying a portion of the weight on the center pin, the bridge turns more easily; meanwhile, the center pin is better held in position, which in turn (because of the connecting spider rods) keeps the rollers aligned in their proper circle.

**riparian rights**   The right of one owning land on or near a body of water to have access to and use of the shore and water.

**rivet**   A metal fastener that was the preferred means of fastening in bridge construction prior to 1970 that is made with a rounded, preformed head at one end and installed while hot into a predrilled or punched hole. The other end is hammered into a similarly shaped head, and the adjoining parts are clamped together.

**rolling-lift bridge**   Patented by William Scherzer in 1893, this bascule bridge operates like a rocking chair by rolling bridge leaves up, back, and away from the waterway to open the draw.

**rusticated masonry**   An architectural feature that contrasts texture with smoothly finished masonry surfaces. Rusticated masonry is usually squared off, but has a rough or uncut outer surface often used for contrast with the smooth masonry above.

**Sanitary and Ship Canal**   *See* Chicago Sanitary and Ship Canal.

**Sanitary District**   *See* Metropolitan Water Reclamation District.

**scow**   Any of various vessels having a flat-bottomed rectangular hull with sloping ends, built in various sizes with or without means of propulsion, as barges, punts, rowboats, or sailboats. Often used as work barges due to their great capacity and large open platform.

**semaphores**   One of the earliest forms of railway signaling, patented in the 1840s. These signals provide indications to train drivers of conditions ahead or warnings for upcoming switches, bridges, or traffic. The designs have altered over the years, and color light signals have mostly replaced semaphore signals. In Chicago semaphores were integral to early bridge operations, as they kept both approaching trains and bridge tenders aware of bridge and train operations, respectively.

**Shuman pavement**   *See* A. F. Shuman pavement.

**single leaf**   Refers to a bascule bridge with a single leaf that spans the waterway to form a continuous bridge deck.

**skew**   The acute angle between the alignment of the superstructure and the alignment of the substructure. When the superstructure is not perpendicular to the substructure, a skew angle is created.

**Slip-Not**   *See* A. F. Shuman pavement.

**span**   The horizontal space between two structural supports. Also refers to the structure itself. May be used as a noun or a verb.

**stay**   Diagonal brace installed to minimize structural movement.

**stringer**   A beam supporting the deck aligned with the length of a span.

**substructure**   The substructure consists of all parts that support the superstructure. The main components are abutments, or end bents; piers, or interior bents; footings; and pilings.

**superstructure**   The superstructure consists of the components that actually span the obstacle the bridge is intended to cross. It includes the bridge deck, trusses, structural members, parapets, handrails, sidewalks, lighting, and drainage features.

**suspension bridge**   A bridge that carries its deck with many tension members, attached to cables and draped over tower piers.

**swing bridge**   A bridge with a moveable deck that opens by rotating horizontally on an axis, usually incorporating a turntable.

**tamarack**   An American larch, *Larix laricina,* of the pine family. The tree has a reddish brown bark and tight clusters of blue-green needles and yields useful timber. The wood from this tree is tough and durable but also flexible in thin strips and was used by the Algonquin people for making snowshoes and in Chicago as paving-block material on the streets and bridges.

**tension**   A stretching force that pulls on a material.

**thru-truss**   A truss that has cross-bracing above and below the roadway and carries traffic through the structure.

**trestle**   An open-braced framework consisting of a succession of towers of steel, timber, or reinforced concrete, supporting short spans. A trestle is a viaduct, but a viaduct is not necessarily a trestle, and the terms are not interchangeable. *See also* viaduct.

**trunnion**   A pair of coaxial projections attached to opposite sides of a cannon, container, bridge leaf, and so on to provide a support about which a bridge leaf can turn in a vertical manner.

**truss**   Any of various structural frames, generally identified by name, based on the geometric form composed of straight members subject to longitudinal compression, tension, or both and further distinguished by the location of the bridge deck in relation to the top and bottom chords. Geometric forms are usually named after their inventors, such as Howe, Pratt, or Warren truss bridges. In relation to deck configuration, traffic traveling on top of the main structure is a deck truss; in a pony configuration, traffic travels between parallel superstructures, which are not cross-braced at the top; in a thru-truss configuration, traffic travels through the superstructure, which is cross-braced above and below traffic.

**U.S. Army Corps of Engineers**   A federal agency and a major army command with the mission of providing vital public engineering services in times of war and peace that strengthen national security, energize the economy, and reduce risks from disasters. The Corps was originally established by President Thomas Jefferson on March 16, 1802. Army engineers began building the Chicago Harbor in 1833, and from 1844 to 1914 the Corps of Engineers constructed harbors and conducted harbor improvements along the Illinois, Indiana, and Wisconsin shorelines.

**vertical-lift bridge**   A moveable deck bridge in which the deck may be raised vertically by synchronized machinery at each end.

**viaduct**   A multispan structure commonly used to carry motor vehicles or rail traffic over a valley, river, roadway, or railroad. *See also* trestle.

**wale**   A horizontal timber or other support used to reinforce various upright members such as pilings. Wales were often used in swing-bridge protection piers to fend off ships.

**warp**   To move (a vessel) into a desired place or position by hauling on a rope that has been fastened to something fixed such as a wharf, piling, buoy, or anchor. Also known as hand-hauling.

**wheel guard**   A raised curb along the outside edge of a traffic lane that separates roadways from sidewalks to safeguard pedestrians on Chicago bridges. These are frequently used to protect construction crews from collisions with vehicles.

**Wolf Point**   A spit of land extending into the Chicago River from the west bank where the North and South branches meet; also, the site of Wolf Point Tavern, the first licensed tavern in Chicago. The spit has since been removed and dredged to allow for more room for the turning of ships. The name is now used to refer to the northeast corner of the North Side where the North Branch and Main Channel meet and current site of the *Chicago Sun-Times* and Merchandise Mart building.

# BIBLIOGRAPHY

Andreas, Alfred T. *History of Chicago: From the Earliest Period to the Present Time.* Vols. 1–3. Chicago: A. T. Andreas, 1884–88.

"Bascule Bridges for the Chicago River." *Engineering News* (November 2, 1899): 286–87.

Becker, Donald N. "Development of the Chicago Type Bascule Bridge." *Proceedings of the American Society of Civil Engineers,* no. 2226 (February 1943): 995–1046.

Bernardoni, Andrea, Mario Taddei, and Edoardo Zanon, eds. *I Ponti di Leonardo* [Leonardo's bridges]. Milan: L3, 2005.

Blanchard, Rufus, *Discovery and Conquests of the Northwest with the History of Chicago.* Vol. 1. Chicago: R. Blanchard, 1898.

Board of Trustees of the Sanitary District of Chicago. *Proceedings of the Board of Trustees of the Sanitary District of Chicago.* Chicago: John F. Higgins, 1898–1920.

Burnham, Daniel H., and Edward H. Bennett. *Plan of Chicago.* New York: Princeton Architectural Press, 1993.

"Cable-Operated Pontoon Bridge over Chicago River." *Engineering News* 93, no. 22 (1924): 864–65.

Chicago Bureau of Engineering, Division of Bridges and Viaducts. *Chicago River Survey.* Chicago: Chicago Bureau of Engineering, November 1953.

Citizens Association of Chicago, Committee on Street Railways and Bridges. *Report of the Committees on Bridges and Street Railways of the Citizens' Association of Chicago.* Chicago: Geo. K. Hazlitt, 1884.

City of Chicago. "Order to the Superintendent to Purchase and Drive Piles to Protect the Lake St. Bridge." *Common Council Proceedings.* Document 53/54 0138 A 04/04, April 4, 1853.

——. "Petition for a Bridge at Lake St." *Common Council Proceedings.* Document 36/36 0238 A 01/05, January 5, 1836.

——. "Proposal of J. F. Brown to Build a Bridge across the South Branch." *Common Council Proceedings.* Document 36/36 0350 A 02/17, February 17, 1836.

——. "Report of Committee on Harbor & Bridges on Petition of A. Crosby and Co., et al. to Widen the Draw of the North Branch Bridge." *Common Council Proceedings.* Document 50/50 6285 A 07/19, July 19, 1850.

——. "Report of Committee on Harbor and Bridges on Petition of S. Lind, et al., to Operate the Lake St. Bridge at Night." *Common Council Proceedings.* Document 53/54 0814 A 11/10, November 11, 1852.

——. "Report of Committee on Harbor & Bridges on Petition of John Rodgers, et al., for a New Bridge at Kinzie St." *Common Council Proceedings.* Document 53/54 0121 A 03/28, March 28, 1853.

——. "Report of Committee on Harbor & Bridges on Petition of Henry Smith, et al., to Erect a Floating Bridge at North Ave." *Common Council Proceedings.* Document 58/59 0456 A 08/02, August 2, 1858.

——. "Report of Committee on Streets and Bridges on the South Branch Bridge." *Common Council Proceedings.* Document 36/36 0350 A 05/11, May 11, 1837.

——. "Report of Special Committee on the Randolph St. Bridge

Announcing Its Completion." *Common Council Proceedings.* Document 50/50 5891 A 01/28, January 28, 1850.

City of Chicago, Board of Public Works. *Fourth Annual Report of the Board of Public Works to the Common Council of the City of Chicago, for the Municipal Fiscal Year Ending April 1, 1865.* Chicago: George H. Fergus, 1865.

——. *Eleventh Annual Report of the Board of Public Works, for the Municipal Year Ending March 31, 1872.* Chicago: National Printing, 1873.

——. *Thirteenth Annual Report of the Board of Public Works.* Chicago: J. S. Thompson & Co. Book and Job Printers, 1874.

City of Chicago, Department of Public Works. *Mayor's Annual Message and the Annual Reports of the Department of Public Works.* Chicago, municipal fiscal years ending December 31, 1876–1969.

——. *Report on Necessary Bridge Work.* Chicago: City of Chicago, Department of Public Works, 1930.

——. *First Annual Report of the Department of Public Works, for the Fiscal Year Ending December 31, 1876.* Chicago: Clark & Edwards, 1877.

——. *Third Annual Report of the Department of Public Works for the Year Ending December 31, 1878.* Chicago: Clark & Edwards, 1879.

——. *Tenth Annual Report of the Department of Public Works, Fiscal Year Ending December 31, 1885.* Chicago: Cameron, Amberg, 1886.

——. *Eleventh Annual Report of the Department of Public Works, for the Year Ending December 31, 1886.* Chicago: M. B. Kenny, 1887.

——. *Thirteenth Annual Report of the Department of Public Works, for the Fiscal Year Ending December 31, 1888.* Chicago: Cameron, Amberg, 1889.

——. *Eighteenth Annual Report of the Department of Public Works, Fiscal Year Ending December 31, 1893.* Chicago: Cameron, Amberg, 1894.

——. *Twenty-First Annual Report of the Department of Public Works for the Year Ending December 31, 1896.* Chicago: Cameron, Amberg, 1897.

——. *Mayor's Annual Message and Thirtieth Annual Report of the Department of Public Works for the Year Ending December 31, 1905.* Chicago: W. J. Hartman, 1906.

——. *Mayor's Annual Message and Thirty-Fifth Annual Report of the Department of Public Works to the Ctiy Council of the City of Chicago for the Year Ending December 31, 1910.* Chicago: City of Chicago, Department of Public Works, 1911.

——. *Fiftieth Annual Report of the Department of Public Works for the Fiscal Year Ending December 31, 1925.* Chicago: John F. Higgins Printers and Binders, 1925.

——. *Fifty-Fourth Annual Report of the Department of Public Works for the Year Ending December 31, 1929.* Chicago, 1930.

City of Chicago, Department of Public Works, Bureau of Engineering, Division of Bridges and Viaducts. *Chicago River Survey, Prepared* in Connection with Pending Application for Federal Permit to Construct a Fixed Bridge over the North Branch of the Chicago River at W. Fullerton Avenue, Specifically Covering the River from W. North Avenue to W. Addison Street. Chicago: City of Chicago, Department of Public Works, November 1953.

——. "List of City Bridges 1914." Courtesy of David Solzman from the estate and private collection of Richard Sutphin. Chicago, 1914.

——. "New South Canal Street Bridge over South Branch Chicago River." Harold Washington Library, Municipal Collection, City of Chicago Public Libraries, Chicago, April 8, 1948.

City of Chicago, Department of Public Works, Division of Bridges and Viaducts. "History of the Washington Street Tunnel under South Branch Chicago River." Harold Washington Library, Municipal Collection, City of Chicago Public Libraries, Chicago, March 30, 1954.

City of Chicago, Department of Transportation, Division of Bridges and Viaducts. "N. Halsted Street (River) Bridge: Statistics and Data." City engineer Stephen J. Michuda, December 1, 1955.

——. "North State Street Bridge and Viaduct." Chicago, January 15, 1942.

Cleaver, Charles, Esq. "Reminiscences of Early Chicago (1833)." Fergus Historical Series 19 (1882).

Condit, Carl W. American Building Art: 19th Century. New York: Oxford University Press, 1960.

——. American Building Art: 20th Century. New York: Oxford University Press, 1961.

Cronon, William. Nature's Metropolis. New York: W. W. Norton, 1991.

Danckers, Ulrich, and Jane Meredith. Early Chicago: A Compendium of the Early History of Chicago to the Year 1835 When the Indians Left. Menomonee Falls, Wis.: Inland Press, 2000.

Gale, Edwin O. Reminiscences of Early Chicago and Vicinity. Chicago: Fleming H. Revell, 1902.

Grossman, James R., Ann Durkin Keating, and Janice L. Reiff, eds. Encyclopedia of Chicago. Chicago: University of Chicago Press, 2004.

Hill, Libby. The Chicago River. Chicago: Lake Claremont Press, 2000.

Hool, George A., and W. S. Kinne. Movable and Long-Span Steel Bridges. New York: McGraw-Hill, 1943.

Hopson, Michael K. The Comprehensive Guide to the Bridges of the Chicago River. Chicago: University of Illinois at Chicago, 1994.

Hubbard, Gurdon S. Incidents and Events in the Life of Gurdon Saltonstall Hubbard. New York: Citadel Press, 1969.

Johnson, John B., Charles Walter Bryan, and Frederick Eugene Turneaure. The Theory and Practice of Modern Framed Structures. New York: John Wiley & Sons, 1903.

Keystone Bridge Company. Descriptive Catalogue of Wrought-Iron Bridges. Philadelphia: Allen, Lane & Scott, 1874.

Larson, John W. Those Army Engineers: A History of the Chicago District, U.S. Army Corps of Engineers. Washington, D.C.: U.S. Government Printing Office, 1980.

Mansfield, J. B., ed. History of the Great Lakes. Vol. 1. Chicago: J. H. Beers, 1899.

Miller, Donald L. City of the Century. New York: Simon & Schuster, 1996.

"The Removal of a Bridge." Engineering News 50, no. 23 (1903): 510.

Simpson, Dick. Rogues, Rebels, and Rubber Stamps. Boulder, Colo.: Westview Press, 2001.

Solzman, David M. The Chicago River. Chicago: University of Chicago Press, 1998.

U.S. Department of the Interior. Historic American Engineering Record. "Addendum to Chicago River Bascule Bridge, Franklin-Orleans Street." Survey no. HAER IL-65. Historian Matthew T. Sneddon, June 1999.

——. Historic American Engineering Record. "Addendum to Chicago River Bascule Bridge, Monroe Street." Survey no. HAER IL-53. Historians Charles Scott, Frances Alexander, and John Nicolay, 1986.

——. Historic American Engineering Record. "Addendum to Chicago River Bascule Bridge, Wells Street Bridge." Survey no. HAER IL-52. Historians Charles Scott, Frances Alexander, and John Nicolay, 1986; Matthew T. Sneddon, 1999.

——. *Historic American Engineering Record.* "Addendum to Chicago River Bascule Bridges." Survey no. HAER IL-111. Historians Francis Alexander, Charles Scott, and John Nicolay, 1986.

——. *Historic American Engineering Record.* "Chicago Avenue Bridge." Survey no. HAER IL-144. Historian Matthew T. Sneddon, June 1999.

——. *Historic American Engineering Record.* "Chicago River Bascule Bridge, Grand Avenue." Survey no. HAER IL-139. Historian: Gary Fitzsimons, 1987.

——. *Historic American Engineering Record.* "Chicago River Bascule Bridge, Wabash Avenue." Survey no. HAER IL-48. Historians Charles Scott, Frances Alexander, and John Nicolay, 1986; Carolyn Brucken, 1992.

——. *Historic American Engineering Record.* "North Avenue Bridge." Survey no. HAER IL-154. Historian Jeffery A. Hess, August 1999.

——. *Historic American Engineering Record.* "North Halsted Street Bridge." Survey no. HAER IL-160. Historian Jeffrey A. Hess, August 1999.

——. *Historic American Engineering Record.* "West Division Street Bridge (Division Street River Bridge)." Survey no. HAER IL-147. Historian Jeffery A. Hess, August 1999.

Vierling, Phillip E. *Early Water Powered Mills of Cook & Lake Counties.* Chicago: Illinois Country Outdoor Guides/Dandellis Printing, 1995.

Wolf, Donald E. *Crossing the Hudson.* New Brunswick, N.J.: Rutgers University Press, 2010.

# INDEX

Note: Page numbers in italics indicate images.

PATRICK T. McBRIARTY is a writer and creative producer based in Chicago. A successful businessperson, he worked many years in manufacturing, including partnership in a small manufacturing company and SAP consulting. He has co-produced, with Stephen Hatch, the documentary film *Chicago Drawbridges.*

The University of Illinois Press
is a founding member of the
Association of American University Presses.

———————————————————

Designed by Dustin Hubbart
and Jim Proefrock
Composed in 10.5/17 Helvetica Neue 55
at the University of Illinois Press
Manufactured by Bang Printing

University of Illinois Press
1325 South Oak Street
Champaign, IL 61820-6903
www.press.uillinois.edu